Selected Papers from ASEPFPM2015

Special Issue Editors

Guanghui Ma
Haruma Kawaguchi
To Ngai

Guest Editors
Guanghui Ma
State Key Laboratory of Biochemical Engineering
China

Haruma Kawaguchi
Kanagawa University
Japan

To Ngai
The Chinese University of Hong Kong
China

Editorial Office	*Publisher*	*Assistant Managing Editor*
MDPI AG	Shu-Kun Lin	Lynn Huang
St. Alban-Anlage 66		
Basel, Switzerland		

This book is a reprint of the Special Issue that appeared in the online, open access journal, *Polymers* (ISSN 2073-4360) in 2016 (available at: http://www.mdpi.com/journal/polymers/special_issues/asepfpm2015)

For citation purposes, cite each article independently as indicated on the article page online and as indicated below:

Author 1; Author 2; Author 3 etc. Article title. *Journal Name*. **Year**. Article number/page range.

ISBN 978-3-03842-314-0 (Pbk)
ISBN 978-3-03842-315-7 (electronic)

Table of Contents

About the Guest Editors

Guang-Hui Ma received her Ph.D. in Polymer Science from the Tokyo Institute of Technology in 1993. She is currently the Leader of the State Key Laboratory of Biochemical Engineering, Institute of Process Engineering, Chinese Academy of Sciences. Prof. Ma's research interests are in the area of nano-/micro-particles preparation and application. Her recent projects include the development of functional nano-/micro-particles as drug delivery systems and vaccine adjuvants. Prof. Ma has published more than 150 SCI articles in peer-reviewed scientific journals, such as *JACS, Adv. Mater., Adv. Funct. Mater., Biomaterials, Nanomedicine,* and *J. Control. Release,* which have been cited overall more than 1100 times. Six English and three Chinese scientific books have also been published and 70 patents (25 granted) and four PCT have been applied for. Prof. Ma was the winner of: the Outstanding Young Scientists Fund (2002); China's State Organization's Award "Women Jiangong Advanced Individual" (2005); "Top ten' Heroine" of CAS (2005); Beijing Award of Science and Technology 1st Prize (2006); National Award for Technological Invention 2nd Prize (2009); and the Asian Young Women Researcher Awards 1st Prize in Engineering and Technology (2009).

Haruma Kawaguchi, received his B.S., M.S. and Ph.D. degrees from Keio University. The title of his Ph.D. thesis, completed under the supervision of Y. Ohtsuka, was "emulsion polymerization initiated by redox system of poly(ethylene-oxide)-containing emulsifier and ceric ion". He started his academic career in the Department of Applied Chemistry of Keio University in 1969. From March 1978 to April 1979, he worked with Professor Otto Vogl at the University of Massachusetts as a postdoctoral researcher. He was promoted to full professor of Keio University in 1989 and presided over the laboratory of Polymer Chemistry. His research interests comprise particle-forming polymerizations such as emulsion polymerization and precipitation polymerization, functional polymeric microspheres such as affinity latex and environment-sensitive microgels, and the physical chemistry of polymer colloids. He retired from Keio University in 2009 and was given the title of Emeritus Professor of Keio University. In the same year, he was invited to be a Special Professor at the Kanagawa University and worked there for 5 years.

To Ngai received his B.S. in chemistry with first class honors at the Chinese University of Hong Kong (CUHK) in 1999. In 2003, he obtained his Ph.D. in chemistry in the same university under the supervision of Professor Chi Wu. He moved to BASF (Ludwigshafen, Germany) in 2003, as a postdoctoral fellow for two years in Dr. Helmut Auweter and Dr. Sven-Holger Behrens's research group. In July 2005, he went to Professor Timothy P. Lodge's group as a postdoctoral fellow in the University of Minnesota. He joined the Department of Chemistry of the CUHK first as a research assistant professor from 2006–2007, and then was appointed as an assistant professor in January 2008. He was early promoted to tenured associate professor in January 2012. His research interests are in various areas of surface, colloid science and soft materials.

Preface to "Selected Papers from ASEPFPM2015"

In recent years, we have witnessed an increased interest towards emulsion polymerization and functional polymeric microspheres. The creation and manipulation of functional components, surfaces, structures and modality at nanoscale of emulsion and microspheres have demonstrated their great success in accelerating development in biotechnology, environmental science, architecture, biomedicine, electronics, food, cosmetics, etc.

While much progress has been made over the past few decades, there are still big challenges for the preparation of functional particles and the verification of preparation mechanisms; there is further room for growth in our ability to provide a high performance in these areas. To fill the unmet needs, the 5th Asian Symposium (ASEPFPM 2015), together with the China State Key Laboratory of Biochemical Engineering (SKLBE) and other top institutes throughout Asian, ushered in the era of emulsion and microspheres to boost their development in bio-manufacture, bio-separation and biomedical engineering. We have established various unique techniques such as membrane emulsification, multi-shell formation, nano-crystallization and a self-assembly approach to prepare nano- and micro- particles of uniform-size and controllable structures. These nano- and micro-particles have proven indispensable in fundamental research as well as in industrial applications on enzyme immobilization, high-throughput protein separation, controlled release, targeted diagnoses, delivery of antigens, siRNA and anti-cancer drugs.

The general theme of this symposium is to discuss recent advances in both fundamentals and applications, as well as to promote science and adaptation in industryl. This Special Issue highlights the advances of emulsion and particles in a wide range of themes, including biofuel and bio-inspired energy devices, targeted drug delivery, tissue engineering, stimuli-sensitive biosensors, theranostics, and construction of novel nano-structures, which hold great promise in industrial applications. This Special Issue is published to provide a platform for other academic and industrial scientists and experts from Asian countries to share their findings, so as to promote potential collaborations. We trust that this Issue can be extended more broadly to applications of functional polymer microspheres by exchanging new ideas and needs, especially in new emerging applications.

We are honored to guest edit this Special Issue and thank all authors and reviewers for their excellent work and unyielding support. Meanwhile, our appreciation goes to the Polymers editorial team for their enthusiastic dedication and professional editing. Contributions from our colleagues at SKLBE and other institutes and corporations that have collaborated are also sincerely appreciated.

Emulsion and microspheres has opened up whole new dimensions in manufacture, agriculture and medical intervention, but wide application of them still requires much effort to be made. By searching for multi-disciplinary techniques and strengthening cooperation between academics and industrialists across the world, we will continue to exploit the full potential of emulsion and microspheres in order to better heal, fuel and feed the world!

Guanghui Ma, Haruma Kawaguchi and To Ngai
Guest Editors

MDPI

Article

Aerosol-Assisted Fast Formulating Uniform Pharmaceutical Polymer Microparticles with Variable Properties toward pH-Sensitive Controlled Drug Release

Hong Lei, Xingmin Gao, Winston Duo Wu *, Zhangxiong Wu * and Xiao Dong Chen

Suzhou Key Laboratory of Green Chemical Engineering, School of Chemical and Environmental Engineering, College of Chemistry, Chemical Engineering and Materials Science, Soochow University, Suzhou 215123, China; 20134209238@stu.suda.edu.cn (H.L.); 20134209233@stu.suda.edu.cn (X.G.); xdchen@mail.suda.edu.cn (X.D.C.)
* Correspondence: duo.wu@suda.edu.cn (W.D.W.); zhangwu@suda.edu.cn (Z.W.);
  Tel.: +86-512-6588-2762 (W.D.W.); +86-512-6588-2782 (Z.W.)

Academic Editor: Guanghui Ma
Received: 6 April 2016; Accepted: 9 May 2016; Published: 14 May 2016

Abstract: Microencapsulation is highly attractive for oral drug delivery. Microparticles are a common form of drug carrier for this purpose. There is still a high demand on efficient methods to fabricate microparticles with uniform sizes and well-controlled particle properties. In this paper, uniform hydroxypropyl methylcellulose phthalate (HPMCP)-based pharmaceutical microparticles loaded with either hydrophobic or hydrophilic model drugs have been directly formulated by using a unique aerosol technique, *i.e.*, the microfluidic spray drying technology. A series of microparticles of controllable particle sizes, shapes, and structures are fabricated by tuning the solvent composition and drying temperature. It is found that a more volatile solvent and a higher drying temperature can result in fast evaporation rates to form microparticles of larger lateral size, more irregular shape, and denser matrix. The nature of the model drugs also plays an important role in determining particle properties. The drug release behaviors of the pharmaceutical microparticles are dependent on their structural properties and the nature of a specific drug, as well as sensitive to the pH value of the release medium. Most importantly, drugs in the microparticles obtained by using a more volatile solvent or a higher drying temperature can be well protected from degradation in harsh simulated gastric fluids due to the dense structures of the microparticles, while they can be fast-released in simulated intestinal fluids through particle dissolution. These pharmaceutical microparticles are potentially useful for site-specific (enteric) delivery of orally-administered drugs.

Keywords: aerosol method; spray drying; polymer microparticle; microencapsulation; drug delivery

1. Introduction

Microencapsulation is highly important for oral and transdermal drug deliveries [1–5]. In particular, microparticles as drug carriers capable of pH-sensitive controlled release play a vital role in site-selective (such as enteric) delivery of orally-administered drugs for therapeutic applications [6–11]. Ideally, the embedded drugs should be well protected by the carrier materials from degradation in the highly-acidic gastric fluid and released in a desirable way while reaching the intestinal tract for efficient absorption [12,13]. Therefore, the selection and control of the structural properties of microparticle carriers are very important [14,15]. Hydroxypropyl methylcellulose phthalate (HPMCP), a generally-recognized-as-safe (GRAS) cellulose derivative, is widely adopted as an enteric polymer for drug delivery due to its pH-dependent solubility, *i.e.*, low solubility in the harsh gastric fluid while the opposite in the intestinal tract [12,16,17]. However, low solubility does not necessarily guarantee drug protection. The structural properties, especially the particle size/uniformity and surface/internal porosity, should also be well controlled to make the encapsulated drugs well protected and to understand release behaviors [2,3,18,19]. Additionally, drug-carrier interactions should be considered for better drug protection and control of particle properties [2].

To fabricate microparticles for the purpose of controlled drug delivery, a series of approaches have been developed, such as the emulsion method [10,20–22], microfluidics [20,23], electrostatic droplet method [24,25], membrane filtration [26], supercritical CO_2 processing [27,28], templating [29], electrohydrodynamic atomization technique [30], and so on. Comparatively, aerosol-based techniques, particularly spray drying technology, is a continuous and low-cost method to manufacture powdered microparticles [31–33]. In addition, microparticles prepared through aerosol methods are usually of high chemical utilization efficiency (theoretically 100%) through avoiding washing and purification steps. In aerosol methods, liquids are atomized to droplets, which are subsequently dried and/or thermally treated to obtain targeted particles [34–37]. During fast solvent evaporation, solutes in liquid droplets are driven away far from equilibrium and forced to be assembled together, leading to unique co-assembly mechanisms and boundary phenomena, and then to novel materials with unique structures, such as mesoporous, core-shell, and hierarchical structures [34–41].

Conventional aerosol techniques, such as conventional spray drying, however, often produce microparticles with relatively small sizes (typically tens of nanometers to a few micrometers) and very broad particle size distributions [34–42], mainly due to the atomized droplets of a wide range of size distributions and complex travelling trajectories, which experience different drying histories within the same product batch. Consequently, the effect of a particular process parameter on microparticle properties and a particular property on particle performance cannot be accurately figured out. To tackle such drawbacks, we have recently introduced a specially-designed micro-fluidic-jet spray dryer (MFJSD) that is capable of producing uniform and large-sized microparticles ranging from 5 to 300 μm with controllable characteristics and functionalities [43–45]. The properties of spray-dried microparticles are significantly influenced by precursor formulation, such as composition and solvent type, and spray drying conditions, such as air flow rate and drying temperature. Therefore, in order to produce pharmaceutical microparticles capable of pH-sensitive drug release, the structures of the microparticles should be precisely controlled and the correlations between particle structure and drug release behavior should be understood.

Herein in this work, for the first time, we explored the use of MFJSD for the one-step straightforward fabrication and control of structural properties of drug-loaded HPMCP-based microparticles for pH-sensitive drug delivery. Hydrocortisone and lysine were selected as a hydrophobic and a hydrophilic model drug, respectively, to investigate their influences on particle properties and release behaviors. The effects of precursor formulation, including solvent composition and drying temperature, on spray dried particle characteristics and release behaviors have been studied in detail. A series of uniform HPMCP-based pharmaceutical microparticles capable of pH-sensitive drug delivery have been successfully fabricated, and their drug release behaviors have been well correlated with their structural properties.

2. Materials and Methods

2.1. Chemicals

Hydroxypropyl methylcellulose phthalate (HPMCP) was purchased from Yolne (Shanghai, China). Hydrocortisone of biochemical reagent was purchased from Sinopharm Chemical Reagent Co., Ltd. (Shanghai, China). Lysine was kindly provided by COFCO (Beijing, China). Ethanol, acetone, hydrochloric acid, ammonium hydroxide solution (25~28 wt %), dimethyl sulfoxide, ninhydrin, and lithium hydroxide monohydrate were of analytical grade. Deionized water (Milli-Q) was used throughout the experiments wherever required.

2.2. Precursor Preparation

The precursors for spray drying were prepared by directly dispersing HPMCP into certain solvents. To investigate the effects of solvents, different formulations of precursors were prepared (Table S1). The total mass content of excipients was set to be 2.5% (w/v). A certain amount, typically 2.5 wt %, of a specific model drug was added into the precursor solutions for one-step formulating pharmaceutical microparticles.

2.3. Microparticle Fabrication

Briefly, the precursor solutions were fed into a standard steel reservoir; and compressed air was used to force the liquid in the reservoir to jet through the microfluidic aerosol nozzle (MFAN, home-made, Suzhou, China, Figure S1a). The liquid flowing rate was controlled by changing the air pressure. The liquid jet was broken-up into droplets by disturbance from vibrating piezoceramics on the nozzle. The vibration frequency was tuned in order to obtain monodisperse droplets. A new generation of MFJSD-6 (home-made, Suzhou, China, Figure S1b) was used to produce microparticles. The inlet temperature was controlled from 95 to 205 °C for various precursor solutions. The flow rate of the hot air was 250 L/min. The collected microparticles were stored in desiccators for further characterizations and drug release tests.

2.4. Particle Characterization

The morphology and structure of microparticles before and after drug release tests were examined by scanning electron microscopy (SEM, S-4700, Hitachi High Technologies Corporation, Tokyo, Japan). Particle size and size distribution were acquired by analyzing SEM images containing over 1000 particles by using Shineso (SHINESO, Hangzhou, China). The average particle size $\bar{d}$ was defined as $\bar{d} = \sum_{i=1}^{n} d_i/N$ and the span of size distribution was described as span $= (d_{90} - d_{10})/d_{50}$, where d_i was the diameter of the i-th particle, N was the total number of micropaticles, and d_{90}, d_{50}, and d_{10} were the cumulative particle sizes at 90%, 50%, and 10%, respectively.

Microparticle density was assessed by filling 0.5~2 g powders into a 5 mL graduated cylinder, and the weight and volume occupied by the powder were recorded. The mass/volume ratio before tapping was calculated as bulk density. The tap density of the powder was then evaluated by tapping the syringe onto a level surface at a height of about 2 cm until no change in volume is observed [46]. Carr's index was calculated as (tap density-bulk density) $\times$ 100/tap density.

The crystalline characteristics of the raw drugs and microparticles were tested by using the powder X-ray diffraction (XRD) technique (X'Pert-Pro MPD, PANalytical B.V., Almelo, Netherlands), with a Cu Kα radiation source and a scan range of 1°/min. The powders were pressed onto quartz blocks by using a glass slide for direct data recording.

2.5. In Vitro Drug Release Test

The swelling and dissolution behaviors of typical pharmaceutical microparticles were studied. Briefly, a small amount of microparticles was placed on a glass slide and a drop of PBS (phosphate buffer solution) or simulated gastric juice (0.1 M HCl) at around 25 °C was dropped on the glass surface next to the microparticles. Then, the microparticles contacted with water and were infiltrated slowly. The dissolution processes in PBS and swelling phenomena in simulated gastric juice were recorded using an optical microscope (OLYMPUS-CX31, Olympus Corporation, Tokyo, Japan) and a series of images at different time intervals were captured by AE (Adobe After Effects CS4, Adobe Systems Incorporated, San Jose, CA, USA).

In a typical drug-release experiment, 25 mL of PBS release medium (pH value 7.4) or simulated gastric fluid (0.1 M HCl solution, pH value 1.0) was transferred into a flask, and the drug-loaded microparticles (~50 mg) were added into the flask. The flask was kept in a shaking incubator at 37 °C with constant agitation (65 rpm). At certain time intervals, 1 mL of the release medium was withdrawn periodically from the flask and replaced with the same amount of fresh release medium. The release medium was passed through a 0.45 µm membrane filter (Millipore, Shanghai, China) in order to remove insoluble matters. The concentration of hydrocortisone was directly measured by UV–Vis spectroscopy (Spectramax M5, Molecular Devices, Silicon Valley, CA, USA) at a wavelength of 247 nm. For the determination of lysine concentration, due to its non-obvious absorption peak in the ultraviolet and visible light areas, a ninhydrin method was adopted for lysine quantification [47,48]. Briefly, samples of releasing medium (0.5 mL) and ninhydrin solution (0.5 mL) were put in tubes and heated in a boiling water bath for 10 min. After heating, the tubes were fully cooled in an ice bath. Then, 2.5 mL of 50% alcohol was added into each tube and thoroughly mixed with a vortex mixer for 15 s. Then, 200 µL of the reaction mixture was transferred into quartz 96-well plates and the lysine concentration was determined by measuring the absorbance values at 570 nm.

3. Results and Discussion

3.1. Particle Formation Process

Solutions composed of HPMCP (Scheme 1a) as the carrier material, hydrophobic hydrocortisone (Scheme 1b) or hydrophilic lysine (Scheme 1c) as the model drug, and ethanol/water mixture as the solvent, were atomized into uniform droplets and directly assembled into uniform microparticles (Scheme 1d–g) via spray drying. At a solid content of 2.5% (w/v), a hydrocortisone-to-HPMCP mass ratio of 0.025:1, and a drying temperature of 155 °C, uniform, discrete, and dimpled spherical microparticles of ~57.9 µm in size (Figure 1a) were obtained by using an ethanol/water mixture of 1:4 volume ratio as the solvent. The surfaces of these microparticles are smooth and dense (Figure 1a inset), indicating the drug molecules are encapsulated inside the HPMCP matrix. The deformation of spherical droplets into dimpled microparticles originates from surface tension gradients of drying droplets. In general, the droplet air/liquid interface recedes while drying (Scheme 1d), and is gradually concentrated by the precursor ingredients that diffuse inward according to Fick's law [49].When the ingredients reach saturation, the droplet surface will be solidified to form a shell and stop shrinking (Scheme 1d,e). On the other hand, drying induces temperature gradients (interior < interface temperature), making surface tension differences (interior > external), and preferential evaporation of ethanol leaves behind more water, further enlarging the surface tension differences, thus creating inward capillary forces (Scheme 1e). The soft nature of the HPMCP polymer (with remaining water solvent) makes the shell unable to withstand the capillary forces, leading to shape deformation (Scheme 1f) and subsequently dimpled or crumpled microparticles (Scheme 1g).

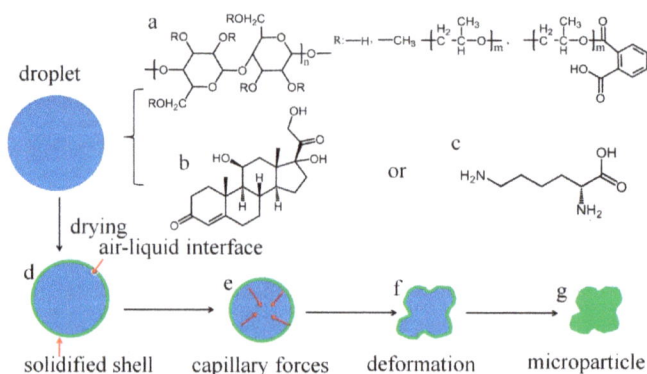

Scheme 1. The molecular structures of the HPMCP polymer (**a**); the model drugs hydrocortisone (**b**); lysine (**c**); and a representation of the formation process (**d**–**g**) of the pharmaceutical microparticles via microfluidic spray drying.

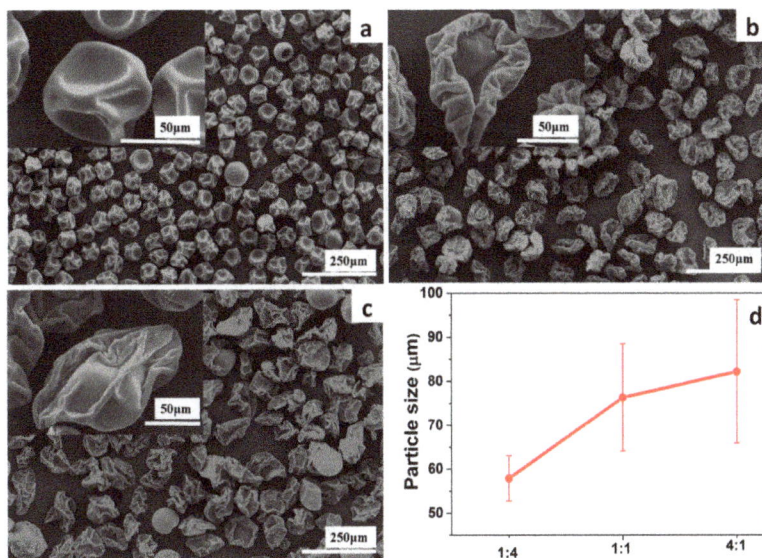

Figure 1. SEM images of the hydrocortisone-loaded microparticles obtained at a drying temperature of 155 °C with an ethanol/water mixed solvent of 1:4 (**a**); 1:1 (**b**); and 4:1 (**c**) volume ratio, respectively; and the particle size variation trend (**d**) with increasing ethanol/water volume ratio.

3.2. Solvent Effect on Particle Property

With other parameters unchanged, by increasing the ethanol/water volume ratio from 1:4 to 1:1, and 4:1, severely crumpled microparticles of more irregular shapes, highly-wrinkled surfaces, and increasing lateral particle sizes (~76.3 and 82.2 µm) were obtained (Figure 1b–d). This solvent effect can be ascribed to the increased evaporation rate by replacing water with more ethanol of higher vapor pressure. With faster solvent removal, the particle shell may form earlier and stabilize the drying droplets, thus leading to increased particle sizes. The fast evaporation rate may also trigger longer paths of receding air/liquid interface and temperature gradient; thus, the capillary forces can drive the drying droplets to crumple more and the surfaces to wrinkle more as well. The solvent effect was further validated by using an acetone/water mixture of 9:1 volume ratio as the solvent. The more volatile acetone solvent further accelerated evaporation, leading to larger-sized and more irregular microparticles, and puffed spherical microparticles of even larger sizes (Figure S2). Under this condition, the solvent evaporation rate is so fast that all of the solvent in a droplet could be evaporated instantaneously to form a relatively rigid shell with negligible remaining water solvent inside, leading to puffed microparticles with smaller degrees of crumpling and wrinkling.

3.3. Drying Temperature Effect on Particle Property

Drug-loaded microparticles were also formulated at different temperatures to further control particle properties. With a constant solid content of 2.5% (w/v), a hydrocortisone-to-HPMCP mass ratio of 0.025:1, and an ethanol/water mixture of 4:1 volume ratio as the solvent, uniform and highly open porous microparticles of ~74.3 µm were obtained at a drying temperature of 95 °C (Figure 2a,b). The microparticles are crumpled in single directions, making them bowl-like without surface wrinkles (Figure 2a,b).The average pore size was estimated to be ~200 nm (Figure 2b and inset), which may accelerate drug release. With the drying temperature slightly increased to 100 °C, the resultant particle morphology changed significantly to severely crumpled and surface-wrinkled microparticles of increased lateral sizes (Figure 2c,f). Interestingly, the microparticles are Janus-like with one half porous, while the other half is non-porous (Figure 2d). With the temperature increased to 155 °C, the obtained microparticles turned to totally non-porous with no obvious particle size increase (Figures 1c and 2f). A further increase of the drying temperature to 205 °C led to puffed microparticles of spherical and smooth shape, as well as an abrupt increase in lateral particle size up to ~133 μm (Figure 2e,f). Very similar trends of temperature-dependent particle properties were also observed for the lysine-loaded microparticles (Figure S3). Similar to the solvent effect, the reason for the particle size increase and higher degree of crumpling/wrinkling with increase of drying temperature is the fastened drying rates at higher temperatures. In addition, at a high drying temperature of 205 °C, the skin formed on the semi-dried droplet may experience higher temperature than the melting point of the HPMCP (~145 °C) for a short time when it travelled along the lower part of the dryer, likely resulting in more smoothness of the particle surface. On the other hand, at a low drying temperature of 95 °C, the slow drying rate allows sufficient solute migration and growth due to the existence of remaining water solvent, thus leading to the formation of bowl-like microparticles with open macropores (Figure 2a,b).

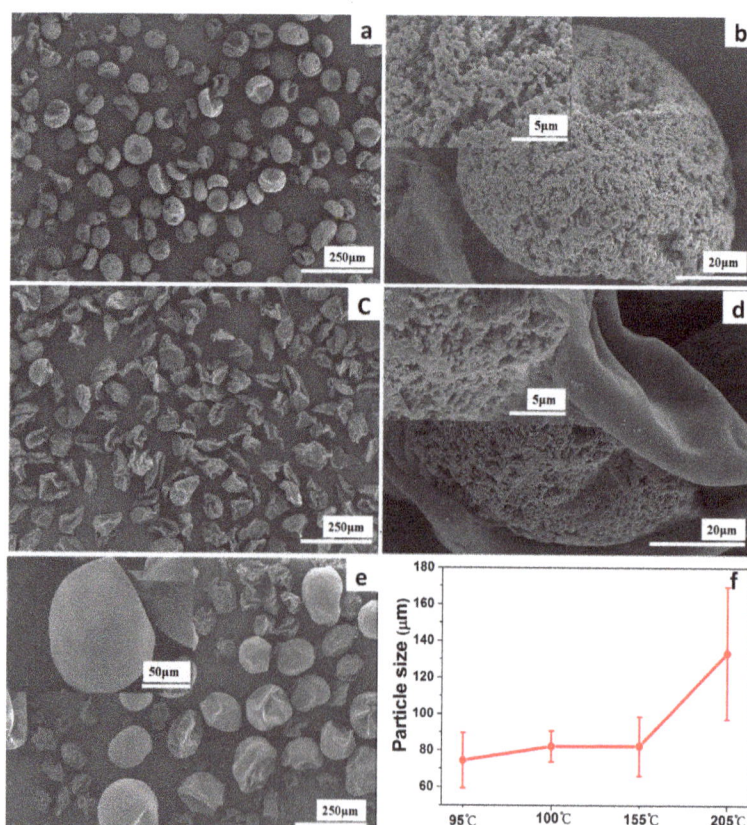

Figure 2. SEM images of the hydrocortisone-loaded microparticles obtained with an ethanol/water mixed solvent of 4:1 volume ratio at a drying temperature of 95 (**a,b**); 100 (**c,d**); and 205 °C (**e**), respectively; and the particle size variation trend (**f**) with increasing drying temperature.

3.4. Effect of Drug Molecule on Particle Properties

Under the same synthetic conditions, the lysine- and hydrocortisone-encapsulated microparticles show very similar properties and variations with solvent composition and drying temperature (Figure 2 and Figure S3). Two differences were observed, though. Firstly, under the same synthetic conditions, the lysine-encapsulated microparticles present a relatively larger particle size than those of the hydrocortisone-encapsulated counterparts (Table S2, for example, comparing the size differences of the hydrocortisone- and lysine-loaded microparticles obtained at 100 or 155 °C). This is probably because of the existence of stronger molecular interactions between lysine (with amino groups, Scheme 1c) and the HPMCP matrix (with carboxylic groups, Scheme 1a), which can accelerate particle solidification during drying. This assumption is verified by the fact that the lateral particle size increases from ~94.6 to 108.8 μm with the increase of the lysine-to-HPMCP mass ratio from 0.025 to 0.1 (Figure S3b,e and Table S2). Secondly, open macroporous bowl-like hydrocortisone-loaded microparticles can be obtained only at a low temperature of 95 °C (Figure 2a), while lysine-loaded microparticles with identical properties can be formed at 100 °C (Figure S3a). This is probably because the hydrophilic lysine molecules can help sustain the water solvent to allow sufficient solute aggregation and growth to form macropores even at relatively higher temperatures.

3.5. Particle Porosity and Density

The porous nature of the bowl-like microparticles was verified by the high surface area (~5.4 m^2/g) of the sample (Figure 3a). With the drying temperature increased to 155 °C or higher, no nitrogen porosity was detected for both the hydrocortisone- and lysine-encapsulated microparticles (Figure 3b,c). At high drying temperature the surface of droplet/particle may be molten, likely leading to a decrease in porosity. There is also a trend that the surface area of the microparticles obtained at the same temperature decreases with the increasing ethanol-to-water volume ratio of the solvents (Figure S4), in accordance with the fact that higher ethanol-to-water volume ratio leads to faster evaporation rates and solidification of particle shells.

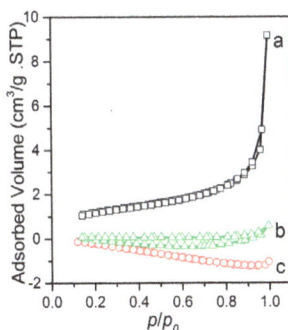

Figure 3. N$_2$ sorption isotherms of the lysine-loaded microparticles obtained with an ethanol/water solvent of 4:1 volume ratio at a drying temperature of 100 (**a**); 155 (**b**); and 205 °C (**c**), respectively.

The flowing property of pharmaceutical microparticles is important for dosing efficiency and product handling, such as mixing, packaging, storage, *etc.* Carr's index (CI) of less than 25% usually indicates fluid powder, which means that the closer bulk and tap densities are, the better the powder flowability. The skeletal density of HPMCP polymer is ~1.25 cm^3/g, while both the bulk and tap densities of the hydrocortisone- and lysine-encapsulated microparticles are far smaller than this value due to the intraparticle porosity and/or inter-particle voids (Table S2). Moreover, both the bulk and tap densities of the obtained microparticles significantly declined with precursor solvent composed of more ethanol, as well as with increased drying temperature (Table S2), much likely resulting from the larger particle size and more irregular shape. The CIs for the hydrocortisone-encapsulated microparticles obtained from different solvent compositions are similar and close to 25%, a sign of good flowability. For the lysine-encapsulated microparticles, the flowability is better for the microparticles obtained at

100 °C than those obtained at higher drying temperatures, probably due to larger and more irregular particle sizes hindering particle compacting.

3.6. States of Drugs in the Microparticles

While the raw drugs are highly crystalline (Figure 4a and Figure S5a), both the encapsulated hydrocortisone and lysine drugs in the microparticles are amorphous regardless of the precursor solvent composition and drying temperature (Figure 4b–d and Figure S5b–d), indicating the drugs are well dispersed in the HCMCP matrix. During the fast spray drying process, the drug molecules are driven far away from equilibrium growth conditions. In addition, the drug molecules could interact strongly with the HPMCP matrix. Therefore, growth and molecular organization of drug molecules are limited, leading to well-dispersed amorphous drug particles. This is believed to be an advantage in terms of facilitating fast drug dissolution and absorption [50].

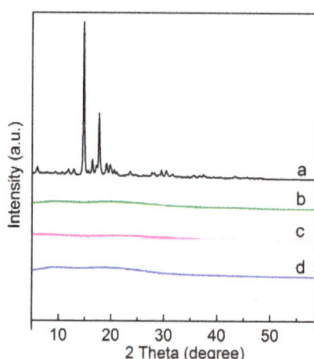

Figure 4. XRD patterns of the raw hydrocortisone drug (a); and the hydrocortisone-loaded microparticles obtained at a drying temperature of 155 °C with an ethanol/water solvent of 1:4 (**b**); 1:1 (**c**); and 4:1 (**d**) volume ratio, respectively.

3.7. Drug Release Behaviors

The release behaviors of the two model drugs were found to be dependent on the structures of the microparticles, as well as sensitive to the pH value of the release environment. In simulated gastric solutions (0.1 M HCl solution), at the same drying temperature (155 °C), for the microparticles obtained at low ethanol/water ratios (1:4 and 1:1), the hydrocortisone drug were released fast with an initial burst of ~70% within one hour (Figure 5a,b). Afterwards, the drug was released slowly with a total of ~80% released up to 6 h (Figure 5a,b), indicating the drugs cannot be well protected in these microparticles against the gastric environment. On the contrary, for the microparticles obtained at a high ethanol/water ratios (4:1), the drug was released very slowly from the beginning (Figure 5c), with only ~20% released within up to 6 h, indicating the drugs in these microparticles can be well protected against the harsh gastric environment.

The above different release behaviors are directly associated with the different structures of the microparticles. The higher ethanol fraction (4:1) in the precursor solvent leads to a faster drying rate, and subsequently to a larger lateral particle size with denser particle shells (Figure 1). As a result, the access of water molecules from the simulated gastric solution into the interior of the microparticles is severely retarded, and the drug molecules are released only very slowly. Tracking of the swelling behaviors of the microparticles in simulated gastric solution further verifies the different release properties. For the microparticles obtained from solvents with low ethanol/water ratios (1:4, 1:1), fast particle swelling and inter-particle mergence were observed (Figure 6a,b), which could trigger fast drug release. On the contrary, for the microparticles obtained from a solvent with the high ethanol/water ratio (4:1), due to the more dense particle shells, only slow and small-degree particle swelling was observed (Figure 6c), which could retard drug release.

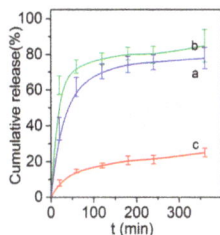

Figure 5. Cumulative release of hydrocortisone drug in a simulated gastric fluid from the microparticles obtained at a drying temperature of 155 °C with an ethanol/water mixed solvent of 1:4 (**a**); 1:1 (**b**); and 4:1 (**c**) volume ratio, respectively.

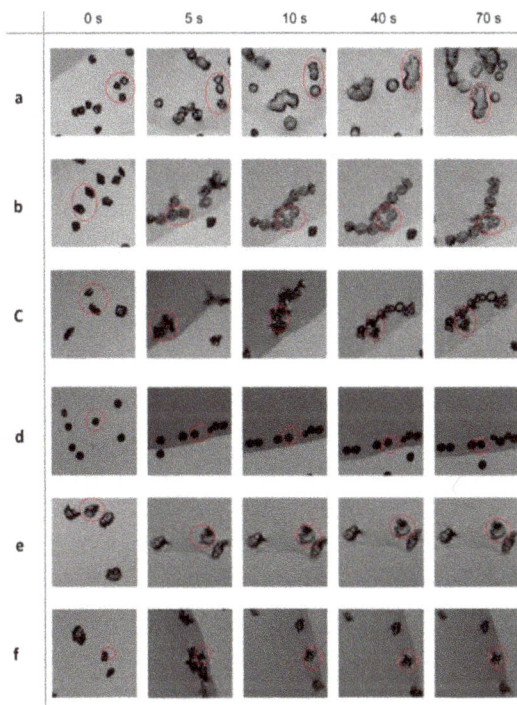

Figure 6. Time-lapse tracking photographs of swelling behaviors in simulate gastric solutions of the microparticles obtained at a drying temperature of 155 °C with an ethanol/water mixed solvent of 1:4 (**a**); 1:1 (**b**); and 4:1 (**c**); volume ratio, and those obtained with an ethanol/water solvent of 4:1 volume ratio at a drying temperature of 95 (**d**); 100 (**e**); and 205 °C (**f**), respectively. A set of the same microparticles in each row was enclosed in red cycles to clearly show their swelling changes with time.

Microparticles obtained at different drying temperatures also show significantly different drug release behaviors. With the same solvent composition (ethanol/water mixture of 4:1 volume ratio), for the microparticles obtained at 95 °C, the encapsulated hydrocortisone drug was released fast with an initial burst of 60%, followed by a gradual release up to ~80% within 6 h (Figure 7a). For those obtained at 100 °C, the release rate became much slower, with an initial release of 25% at 20 min and then a gradual one up to ~40% at 6 h (Figure 7b). Notably, a much slower rate was present for those obtained at a higher temperature of 155 °C (Figure 7c). For the lysine-encapsulated microparticles, very similar temperature-dependent releasing trends were observed (Figure 7d–f). The above different drying-temperature-induced release behaviors can also be well correlated with the structural properties of the corresponding microparticles. At low drying temperatures, the obtained microparticles are open porous (Figure 2a and Figure S3a) so that the access of water into the interior of these microparticles is facilitated, and the drug molecules inside can be released very fast. Tracking of these microparticles reveals that they were not obviously swollen in the simulated gastric solution (Figure 6d), further indicating the drug release mainly relies on direct mass transportation through the macropores. On the contrary, the microparticles obtained at higher temperatures (155 °C or higher) are quite dense (Figure 1) without any detectable N_2 porosity (Figure 3). Besides, at high drying temperatures, the obtained microparticles appear crumpled/wrinkled or puffed morphology with a possibly molten surface, which could restrict drug release. Therefore, mass exchange between outside and internal microparticles is much retarded, leading to much slower drug release rates. Tracking of these microparticles reveals that the microparticles were swollen in the simulated gastric solution (Figure 6e,f), indicating the drug release mainly relies on slow mass exchange.

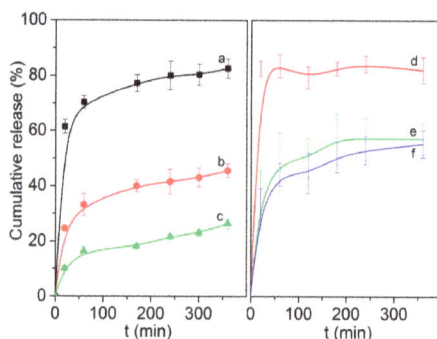

Figure 7. Cumulative release of hydrocortisone (**a–c**); and lysine (**d–f**) in simulated gastric fluids from the microparticles obtained with an ethanol/water mixed solvent of 4:1 volume ratio at a drying temperature of 95 (**a**); 100 (**b,d**); 155 (**c,e**); and 205 °C (**f**), respectively.

Interestingly, under the same synthetic conditions, the lysine-loaded microparticles show faster drug release rates as compared to those of the hydrocortisone-loaded counterparts (Figure 7). For example, for the microparticles obtained at 100 °C, the released amount of lysine was up to 80% within 1 h and without noticeable release afterwards (Figure 7d). However, the released amount of hydrocortisone was only ~32 wt % within 1 h and with a further slow release stage (Figure 7b). The reason is mainly due to the difference in surface pore structure. The former shows highly open porous structure (Figure S3a) while the latter is only half-particle porous and the pores are less open (Figure 2c,d). Similar release difference was also found in the microparticles obtained at 155 °C (Figure 7c,e). At this temperature, both the lysine- and hydrocortisone-loaded microparticles are quite dense; therefore, the faster release of lysine is probably due to its more basic and hydrophilic nature and smaller molecular size as compared with hydrocortisone, which make it more easily to be dissolved in the simulated gastric solution. This is verified by the much faster release rate of the lysine-loaded microparticles obtained at 100 °C (Figure 7d) than that of the hydrocortisone-loaded microparticles obtained at 95 °C (Figure 7a), while both types of microparticles show highly open porous structures (Figure 2a and Figure S3a).

After the release tests in simulated gastric solutions, both the hydrocortisone- and lysine-loaded microparticles show no obvious changes in particle morphology and structure (Figures S6 and S7). These results indicate the microparticles can be well preserved in the harsh simulated gastric environments.

The drug release is pH-sensitive with totally different release behaviors near a neutral environment. To mimic the enteric environment, a PBS solution (pH value 7.4) was used as the release medium. Notably, ~90% drug molecules (either lysine or hydrocortisone) were released in the first 30 min and almost 100% released within 1 h for all the drug-encapsulated microparticles regardless of their structural properties (Figure 8). Track of the microparticles in PBS solution indicates that the release mechanism lies in the fast dissolution of the whole microparticles in PBS solution (Figure 9 and Figure S8), in accordance with literature that HPMCP is a polymer with high solubility in the intestinal tract [51,52]. The above results indicate that site-specific drug delivery could be achieved by controlling the microparticle properties.

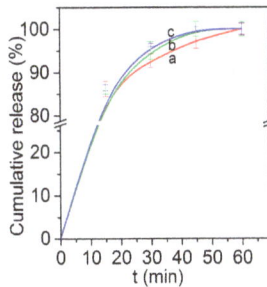

Figure 8. Cumulative release of lysine in PBS solutions from the microparticles obtained with an ethanol/water solvent of 4:1 volume ratio at a drying temperature of 100 (**a**); 155 (**b**); and 205 °C (**c**), respectively.

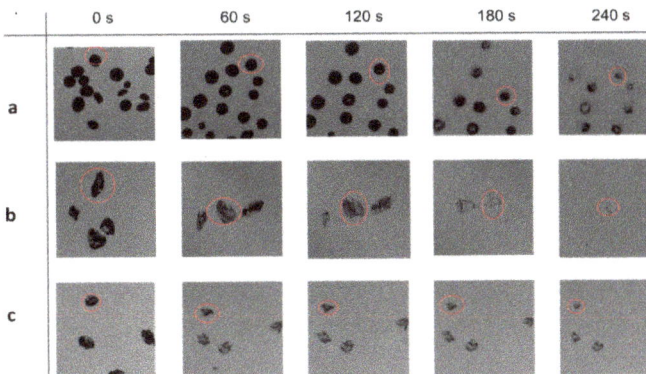

Figure 9. Time-lapse tracking photographs of dissolution behaviors in PBS solutions of the lysine-loaded microparticles obtained with an ethanol/water solvent of 4:1 volume ratio at a drying temperature of 100 (**a**); 155 (**b**); and 205 °C (**c**), respectively.

4. Conclusions

This paper has demonstrated the use of microfluidic jet spray drying to directly fabricate drug-encapsulated HPMCP polymer microparticles with uniform and tunable sizes (57~133 μm), variable morphologies (bowl-like, dimpled, crumpled, and spherical) and structures (porous, Janus-like, porous, and dense). Adoption of either more volatile solvents or higher drying temperatures leads to microparticles with larger lateral particle sizes, more severely crumpled morphologies and wrinkled surfaces, and declined porosity. At high drying temperature the surface of droplet/particle may be molten during spray drying process, also likely leading to more smoothness of the particle surface and decrease in porosity. As compared with the hydrophobic hydrocortisone, the hydrophilic and basic lysine model drug tends to interact more strongly with HMPCP leading to relatively larger particle sizes, to sustain more water leading to the formation of open macropores at a relatively higher drying temperature, and to be released relatively faster. The release behaviors are particle-property dependant and pH-sensitive. Drugs in the microparticles obtained at a drying temperature of 155 °C or higher, and those obtained by using a solvent with a high ethanol/water ratio can be well protected in simulated gastric solutions and subsequently be fast released in PBS solutions. Therefore, these microparticles are capable of protecting drugs against harsh gastric fluids and releasing them in enteric fluids. The current method could provide a robust route for the fabrication for pharmaceutical microparticles for site-specific delivery of orally administrated drugs of different molecular properties.

Supplementary Materials: Supplementary Materials can be found at www.mdpi.com/2073-4360/8/5/195/s1. Figure S1: Photographs of the MFAN nozzle (a); and the spray drying set-up (b); Figure S2: SEM images of the hydrocortisone-loaded microparticles obtained by using an acetone/water solvent of 9:1 volume ratio at a drying temperature of 155 °C; Figure S3: SEM images of the lysine-loaded microparticles obtained with an ethanol/water mixed solvent of 4:1 volume ratio at a drying temperature of 100 (a); 155 (b); and 205 °C (c); respectively, and the particle size variation trend (d) with increasing drying temperature; (e) and (f) are the SEM images of the lysine-loaded microparticles with 10% drug loading obtained with an ethanol/water mixed solvent of 4:1 volume ratio at a drying temperature o155 °C; Figure S4: N_2 sorption isotherms of the lysine-loaded microparticles obtained with an ethanol/water solvent of 1: 4 (a); and 1:1 (b) volume ratio at a drying temperature of 155 °C; Figure S5: XRD patterns of the raw lysine drug (a); and the lysine-loaded microparticles obtained with an ethanol/water mixed solvent of 4:1 at a drying temperature of 100 (b); 155 (c); and 205 °C (d), respectively; Figure S6: After the drug release tests in simulated gastric solutions, SEM images of the hydrocortisone-loaded microparticles obtained at a drying temperature of 155 °C with an ethanol/water mixed solvent of 1:4 (a); 1:1 (b); and 4:1 (c) volume ratio, respectively; Figure S7: After the drug release tests in simulated gastric solutions, SEM images of the lysine-loaded microparticles obtained with an ethanol/water mixed solvent of a volume ratio of 4:1 at a drying temperature of 100 (a); 155 (b); and 205 °C (c), respectively; Figure S8: Time-lapse tracking photographs of dissolution behaviors in PBS solutions of the hydrocortisone-loaded microparticles obtained with an ethanol/water mixed solvent of 1:4 (a); 1:1 (b); and 4:1 (c) volume ratio at a drying temperature of 155 °C. Table S1: Summary of the formulations of precursors for the fabrication of a series of pharmaceutical microparticles via microfluidic jet spray drying; Table S2: Data summary of the particle size, bulk density, tap density, and Carr's Index for the obtained pharmaceutical microparticles.

Acknowledgments: This work was supported by the National Natural Science Foundation of China (No. 21506135 and 21501125), the Natural Science Foundation of Jiangsu Province (BK20140317 and BK20150312), the Young Thousand Talented Program (2015), and the Start-up fund (Q410900115) of Soochow University. The support from the Priority Academic Program Development (PAPD) of Jiangsu Higher Education Institutions was appreciated.

Author Contributions: Hong Lei, Winston Duo Wu and Zhangxiong Wu conceived and designed the experiments; Hong Lei and Xingmin Gao performed the experiments; Hong Lei, Winston Duo Wu and Zhangxiong Wu analyzed the data; Hong Lei, Winston Duo Wu, Zhangxiong Wu and Xiao Dong Chen wrote the paper.

Conflicts of Interest: The authors declare no conflict of interest.

Abbreviations

The following abbreviations are used in this manuscript:

HPMCP	Hydroxypropyl methylcellulose phthalate
GRAS	Generally recognized as safe
MFJSD	Micro-fluidic-jet spray dryer
MFAN	Micro-fluidic-aerosol-nozzle
AE	Adobe after effects CS4
CI	Carr's index
XRD	X-ray diffraction
SEM	Scanning electron microscopy
PBS	Phosphate buffer saline
UV	Ultraviolet

References

1. Mathiowitz, E.; Jacob, J.S.; Jong, Y.S.; Carino, G.P.; Chickering, D.E.; Chaturvedi, P.; Santos, C.A.; Vijayaraghavan, K.; Montgomery, S.; Bassett, M.; *et al.* Biologically erodable microspheres as potential oral drug delivery systems. *Nature* **1997**, *386*, 410–414. [CrossRef] [PubMed]
2. Lam, P.L.; Gambari, R. Advanced progress of microencapsulation technologies: *In vivo* and *in vitro* models for studying oral and transdermal drug deliveries. *J. Controll. Release* **2014**, *178*, 25–45. [CrossRef] [PubMed]
3. Ma, G. Microencapsulation of protein drugs for drug delivery: strategy, preparation, and applications. *J. Controll. Release* **2014**, *193*, 324–340. [CrossRef] [PubMed]
4. Wang, W.; Liu, X.; Xie, Y.; Zhang, H.A.; Yu, W.; Xiong, Y.; Xie, W.; Ma, X. Microencapsulation using natural polysaccharides for drug delivery and cell implantation. *J. Mater. Chem.* **2006**, *16*, 3252–3267. [CrossRef]
5. Johnston, A.P.R.; Such, G.K.; Caruso, F. Triggering release of encapsulated cargo. *Angew. Chem. Int. Ed.* **2010**, *49*, 2664–2666. [CrossRef] [PubMed]
6. Calija, B.; Cekic, N.; Savic, S.; Daniels, R.; Markovic, B.; Milic, J. pH-Sensitive microparticles for oral drug delivery based on alginate/oligochitosan/eudragit® l100-55 "sandwich" polyelectrolyte complex. *Colloids Surf. B Biointerfaces* **2013**, *110*, 395–402. [CrossRef] [PubMed]
7. Jablan, J.; Jug, M. Development of eudragit® s100 based pH-responsive microspheres of zaleplon by spray-drying: tailoring the drug release properties. *Powder Technol.* **2015**, *283*, 334–343. [CrossRef]
8. Raizaday, A.; Yadav, H.K.; Kumar, S.H.; Kasina, S.; Navya, M.; Tashi, C. Development of pH sensitive microparticles of karaya gum: By response surface methodology. *Carbohydr. Polym.* **2015**, *134*, 353–363. [CrossRef] [PubMed]
9. Xiao, B.; Si, X.; Zhang, M.; Merlin, D. Oral administration of pH-sensitive curcumin-loaded microparticles for ulcerative colitis therapy. *Colloids Surf. B Biointerfaces* **2015**, *135*, 379–385. [CrossRef] [PubMed]
10. Arimoto, M.; Ichikawa, H.; Fukumori, Y. Microencapsulation of water-soluble macromolecules with acrylic terpolymers by the wurster coating process for colon-specific drug delivery. *Powder Technol.* **2004**, *141*, 177–186. [CrossRef]
11. De Jaeghere, F.; Allémann, E.; Kubel, F.; Galli, B.; Cozens, R.; Doelker, E.; Gurny, R. Oral bioavailability of a poorly water soluble HIV-1 protease inhibitor incorporated into pH-sensitive particles: Effect of the particle size and nutritional state. *J. Controll. Release* **2000**, *68*, 291–298. [CrossRef]
12. Colombo, S.; Brisander, M.; Haglöf, J.; Sjövall, P.; Andersson, P.; Østergaard, J.; Malmsten, M. Matrix effects in nilotinib formulations with pH-responsive polymer produced by carbon dioxide-mediated precipitation. *Int. J. Pharm.* **2015**, *494*, 205–217. [CrossRef] [PubMed]
13. Kossena, G.A.; Charman, W.N.; Boyd, B.J.; Porter, C.J.H. A novel cubic phase of medium chain lipid origin for the delivery of poorly water soluble drugs. *J. Controll. Release* **2004**, *99*, 217–229. [CrossRef] [PubMed]
14. Malgras, V.; Ji, Q.; Kamachi, Y.; Mori, T.; Shieh, F.K.; Wu, K.C.; Ariga, K.; Yamauchi, Y. Templated synthesis for nanoarchitectured porous materials. *Bull. Chem. Soc. Jpn.* **2015**, *88*, 1171–1200. [CrossRef]
15. Baeza, A.; Colilla, M.; Vallet-Regí, M. Advances in mesoporous silica nanoparticles for targeted stimuli-responsive drug delivery. *Exp. Opin. Drug Deliv.* **2015**, *12*, 319–337. [CrossRef] [PubMed]
16. Kim, I.H.; Park, J.H.; Cheong, I.W.; Kim, J.H. Swelling and drug release behavior of tablets coated with aqueous hydroxypropyl methylcellulose phthalate (HPMCP) nanoparticles. *J. Controll. Release* **2003**, *89*, 225–233. [CrossRef]

17. Singh, B.; Maharjan, S.; Jiang, T.; Kang, S.-K.; Choi, Y.-J.; Cho, C.-S. Attuning hydroxypropyl methylcellulose phthalate to oral delivery vehicle for effective and selective delivery of protein vaccine in ileum. *Biomaterials* **2015**, *59*, 144–159. [CrossRef] [PubMed]

18. Richardson, J.J.; Björnmalm, M.; Caruso, F. Technology-driven layer-by-layer assembly of nanofilms. *Science* **2015**. [CrossRef] [PubMed]

19. Ariga, K.; Yamauchi, Y.; Rydzek, G.; Ji, Q.; Yonamine, Y.; Wu, K.C.W.; Hill, J.P. Layer-by-layer nanoarchitectonics: Invention, innovation, and evolution. *Chem. Lett.* **2014**, *43*, 36–68. [CrossRef]

20. Datta, S.S.; Abbaspourrad, A.; Amstad, E.; Fan, J.; Kim, S.-H.; Romanowsky, M.; Shum, H.C.; Sun, B.; Utada, A.S.; Windbergs, M.; *et al.* Double emulsion templated solid microcapsules: mechanics and controlled release. *Adv. Mater.* **2014**, *26*, 2205–2218. [CrossRef] [PubMed]

21. Barbe, C.; Bartlett, J.; Kong, L.G.; Finnie, K.; Lin, H.Q.; Larkin, M.; Calleja, S.; Bush, A.; Calleja, G. Silica particles: A novel drug-delivery system. *Adv. Mater.* **2004**, *16*, 1959–1966. [CrossRef]

22. Little, S.R.; Lynn, D.M.; Puram, S.V.; Langer, R. Formulation and characterization of poly(β amino ester) microparticles for genetic vaccine delivery. *J. Controll. Release* **2005**, *107*, 449–462. [CrossRef] [PubMed]

23. Dendukuri, D.; Doyle, P.S. The synthesis and assembly of polymeric microparticles using microfluidics. *Adv. Mater.* **2009**, *21*, 4071–4086. [CrossRef]

24. Shang, Y.; Ding, F.; Xiao, L.; Deng, H.; Du, Y.; Shi, X. Chitin-based fast responsive pH sensitive microspheres for controlled drug release. *Carbohydr. Polym.* **2014**, *102*, 413–418. [CrossRef] [PubMed]

25. Fattahi, P.; Borhan, A.; Abidian, M.R. Microencapsulation of chemotherapeutics into monodisperse and tunable biodegradable polymers via electrified liquid jets: Control of size, shape, and drug release. *Adv. Mater.* **2013**, *25*, 4555–4560. [CrossRef] [PubMed]

26. Wei, W.; Yuan, L.; Hu, G.; Wang, L.-Y.; Wu, J.; Hu, X.; Su, Z.-G.; Ma, G.-H. Monodisperse chitosan microspheres with interesting structures for protein drug delivery. *Adv. Mater.* **2008**, *20*, 2292–2296. [CrossRef]

27. Hassani, L.N.; Hindre, F.; Beuvier, T.; Calvignac, B.; Lautram, N.; Gibaud, A.; Boury, F. Lysozyme encapsulation into nanostructured $CaCO_3$ microparticles using a supercritical CO_2 process and comparison with the normal route. *J. Mater. Chem. B* **2013**, *1*, 4011–4019. [CrossRef]

28. Whitaker, M.J.; Hao, J.; Davies, O.R.; Serhatkulu, G.; Stolnik-Trenkic, S.; Howdle, S.M.; Shakesheff, K.M. The production of protein-loaded microparticles by supercritical fluid enhanced mixing and spraying. *J. Control. Release* **2005**, *101*, 85–92. [CrossRef] [PubMed]

29. Kelly, J.Y.; DeSimone, J.M. evidence for partially bound states in cooperative molecular recognition interfaces. *J. Am. Chem. Soc.* **2008**, *130*, 5438–5439. [CrossRef] [PubMed]

30. Seremeta, K.P.; Hocht, C.; Taira, C.; Cortez Tornello, P.R.; Abraham, G.A.; Sosnik, A. Didanosine-loaded poly(epsilon-caprolactone) microparticles by a coaxial electrohydrodynamic atomization (CEHDA) Technique. *J. Mater. Chem. B* **2015**, *3*, 102–111. [CrossRef]

31. Bore, M.T.; Rathod, S.B.; Ward, T.L.; Datye, A.K. Hexagonal mesostructure in powders produced by evaporation-induced self-assembly of aerosols from aqueous tetraethoxysilane solutions. *Langmuir* **2003**, *19*, 256–264. [CrossRef]

32. Esposito, E.; Cervellati, F.; Menegatti, E.; Nastruzzi, C.; Cortesi, R. Spray dried eudragit microparticles as encapsulation devices for vitamin C. *Int. J. Pharm.* **2002**, *242*, 329–334. [CrossRef]

33. Rizi, K.; Green, R.J.; Donaldson, M.; Williams, A.C. Production of pH-responsive microparticles by spray drying: Investigation of experimental parameter effects on morphological and release properties. *J. Pharm. Sci.* **2011**, *100*, 566–579. [CrossRef] [PubMed]

34. Boissiere, C.; Grosso, D.; Chaumonnot, A.; Nicole, L.; Sanchez, C. Aerosol route to functional nanostructured inorganic and hybrid porous materials. *Adv. Mater.* **2011**, *23*, 599–623. [CrossRef] [PubMed]

35. Lu, Y.; Fan, H.; Stump, A.; Ward, T.L.; Rieker, T.; Brinker, C.J. Aerosol-assisted self-assembly of mesostructured spherical nanoparticles. *Nature* **1999**, *398*, 223–226.

36. Tsung, C.-K.; Fan, J.; Zheng, N.; Shi, Q.; Forman, A.J.; Wang, J.; Stucky, G.D. A general route to diverse mesoporous metal oxide submicrospheres with highly crystalline frameworks. *Angew. Chem. Int. Ed.* **2008**, *47*, 8682–8686. [CrossRef] [PubMed]

37. Boissiere, C.; Grosso, D.; Amenitsch, H.; Gibaud, A.; Coupe, A.; Baccile, N.; Sanchez, C. First *in-situ* SAXS studies ofthe mesostructuration of spherical silica and titania particles during spray-drying process. *Chem. Commun.* **2003**, *22*, 2798–2799. [CrossRef]

38. Yan, Y.; Zhang, F.Q.; Meng, Y.; Tu, B.; Zhao, D.Y. One-step synthesis of ordered mesoporous carbonaceous spheres by an aerosol-assisted self-assembly. *Chem. Commun.* **2007**, *27*, 2867–2869. [CrossRef] [PubMed]

39. Suh, W.H.; Kang, J.K.; Suh, Y.H.; Tirrell, M.; Suslick, K.S.; Stucky, G.D. Porous carbon produced in air: Physicochemical propertieand stem cell engineering. *Adv. Mater.* **2011**, *23*, 2332–2338. [CrossRef] [PubMed]

40. Xu, H.; Guo, J.; Suslick, K.S. Porous carbon spheres from energetic carbon precursors using ultrasonic spray pyrolysis. *Adv. Mater.* **2012**, *24*, 6028–6033. [CrossRef] [PubMed]

41. Liu, W.; Wu, W.D.; Selomulya, C.; Chen, X.D. Facile spray-drying assembly of uniform microencapsulates with tunable core shell structures and controlled release properties. *Langmuir* **2011**, *27*, 12910–12915. [CrossRef] [PubMed]

42. Friesen, D.T.; Shanker, R.; Crew, M.; Smithey, D.T.; Curatolo, W.J.; Nightingale, J.A.S. Hydroxypropyl methylcellulose acetate succinate-based spray-dried dispersions: An overview. *Mol. Pharm.* **2008**, *5*, 1003–1019. [CrossRef] [PubMed]

43. Wu, W.D.; Lin, S.X.; Chen, X.D. Monodisperse droplet formation through a continuous jet break-up using glass nozzles operated with piezoelectric pulsation. *AIChE J.* **2011**, *57*, 1386–1392. [CrossRef]

44. Wu, W.D.; Ria, A.; Na, H.; Cordelia, S.; Zhao, D.; Yu-Lung, C.; Dong, C.X. Assembly of uniform photoluminescent microcomposites using a novel micro-fluidic-jet-spray-dryer. *AIChE J.* **2011**, *57*, 2726–2737. [CrossRef]

45. Wu, Z.; Wu, W.D.; Liu, W.; Selomulya, C.; Chen, X.D.; Zhao, D. A general "surface-locking" approach toward fast assembly and processing of large-sized, ordered, mesoporous carbon microspheres. *Angew. Chem. Int. Ed.* **2013**, *52*, 13764–13768. [CrossRef] [PubMed]

46. Healy, A.M.; Mcdonald, B.F.; Tajber, L.; Corrigan, O.I. Characterisation of excipient-free nanoporous microparticles (NPMPs) of BendroFlumethiazide. *Eur. J. Pharm. Biopharm.* **2008**, *69*, 1182–1186. [CrossRef] [PubMed]

47. Moore, S.; Stein, W.H. A family of basic amino acid transporters of the vacuolar membrane from *Saccharomyces cerevisiae*. *J. Biol. Chem.* **2005**, *280*, 4851–4857.

48. Sun, S.W.; Lin, Y.C.; Weng, Y.M.; Chen, M.J. Efficiency improvements on ninhydrin method for amino acid quantification. *J. Food Compos. Anal.* **2006**, *19*, 112–117. [CrossRef]

49. Reinhard, V. Pharmaceutical particle engineering via spray drying. *Pharmacol. Res.* **2008**, *25*, 999–1022.

50. Singhal, D.; Curatolo, W. Drug polymorphism and dosage form design: A practical perspective. *Adv. Drug Deliv. Rev.* **2004**, *56*, 335–347. [CrossRef] [PubMed]

51. Eiamtrakarn, S.; Itoh, Y.; Kishimoto, J.; Yoshikawa, Y.; Shibata, N.; Murakami, M.; Takada, K. Gastrointestinal mucoadhesive patch system (GI-MAPS) for oral administration of G-CSF, a model protein. *Biomaterials* **2002**, *23*, 145–152. [CrossRef]

52. Yang, M.; Cui, F.; You, B.; You, J.; Wang, L.; Zhang, L.; Kawashima, Y. A novel pH-dependent gradient-release delivery system for nitrendipine—I. Manufacturing, evaluation *in vitro* and bioavailability in healthy dogs. *J. Controll. Release* **2004**, *98*, 219–229.

polymers

MDPI

Article

Co-Assembly of Graphene Oxide and Albumin/Photosensitizer Nanohybrids towards Enhanced Photodynamic Therapy

Ruirui Xing [1,2,3], Tifeng Jiao [1,2,*], Yamei Liu [1,2], Kai Ma [1,2], Qianli Zou [3,4,*], Guanghui Ma [3] and Xuehai Yan [3,4,*]

[1] State Key Laboratory of Metastable Materials Science and Technology, Yanshan University, Qinhuangdao 066004, China; rrxing@ipe.ac.cn (R.X.); liuym@ipe.ac.cn (Y.L.); makai@ipe.ac.cn (K.M.)
[2] Hebei Key Laboratory of Applied Chemistry, School of Environmental and Chemical Engineering, Yanshan University, Qinhuangdao 066004, China
[3] National Key Laboratory of Biochemical Engineering, Institute of Process Engineering, Chinese Academy of Sciences, Beijing 100190, China; ghma@ipe.ac.cn
[4] Center for Mesoscience, Institute of Process Engineering, Chinese Academy of Sciences, Beijing 100190, China
* Correspondence: tfjiao@ysu.edu.cn (T.J.); qlzou@ipe.ac.cn (Q.Z.); yanxh@ipe.ac.cn (X.Y.); Tel.: +86-335-806-1569 (T.J.); +86-10-8254-5024 (Q.Z. & X.Y.)

Academic Editors: Jung Kwon Oh and Shiyong Liu
Received: 29 January 2016; Accepted: 27 April 2016; Published: 4 May 2016

Abstract: The inactivation of photosensitizers before they reach the targeted tissues can be an important factor, which limits the efficacy of photodynamic therapy (PDT). Here, we developed co-assembled nanohybrids of graphene oxide (GO) and albumin/photosensitizer that have a potential for protecting the photosensitizers from the environment and releasing them in targeted sites, allowing for an enhanced PDT. The nanohybrids were prepared by loading the pre-assembled nanoparticles of chlorin e6 (Ce6) and bovine serum albumin (BSA) on GO via non-covalent interactions. The protection to Ce6 is evident from the inhibited fluorescence and singlet oxygen generation activities of Ce6–BSA–GO nanohybrids. Importantly, compared to free Ce6 and Ce6 directly loaded by GO (Ce6–GO), Ce6–BSA–GO nanohybrids showed enhanced cellular uptake and *in vitro* release of Ce6, leading to an improved PDT efficiency. These results indicate that the smart photosensitizer delivery system constructed by co-assembly of GO and albumin is promising to improve the stability, biocompatibility, and efficiency of PDT.

Keywords: protein; graphene oxide; singlet oxygen generation; co-assembly; photodynamic therapy

1. Introduction

Photodynamic therapy (PDT) is attractive for cancer therapy due to its specific selectivity to a disease site and non-invasive protocol [1,2]. Essentially, PDT involves the administration of a photoactive drug (photosensitizer), followed by selective irradiation of the cancerous tissue by light [3]. The activated photosensitizer reacts with molecular oxygen, resulting in the formation of reactive oxygen species (ROS), such as singlet oxygen, that are directly responsible for the death of cancer cells [4]. Hence, the cytotoxicity in PDT is a combined result of three nontoxic components (photosensitizer, molecular oxygen, and light), leading PDT to be a highly-biocompatible treatment modality for cancer. Despite the fast-growing research about PDT [5–11], there are still huge challenges before its acceptance as a first-line oncological treatment. Due to the limited selective accumulation of photosensitizers to tumors [12], a large amount of photosensitizers are needed to obtain a satisfactory photodynamic response. Importantly, it typically takes 48–72 h for the photosensitizers to accumulate in tumors, leading to the unprotected photosensitizers vulnerable in the circulatory system [13]. An intriguing method for circumventing these challenges is to formulate the photosensitizers in means of suitable nanocarriers [13–16]. The nanocarriers can be tailored to protect the encapsulated photosensitizers in circulation and readily release them in targeted tissues.

Due to its special photochemical properties, graphene has attracted remarkable attention in many fields including smart drug delivery [17–19]. The planar and highly-conjugated morphology of graphene sheet is ideally suitable for loading of drugs through the non-covalent interactions, such as hydrophobic interaction and π–π stacking [20]. The unique two-dimensional structure of graphene sheet ensures a high loading capability for physical adsorption of drugs on its large surface [21]. Graphene oxide (GO), a derivative of graphene, not only has the same merits as graphene but also holds better solubility and more functional groups suitable for further modification [22,23]. GO can be converted to nano-GO with the lateral dimensions in nanometric size by applying ultrasonic energy [24,25]. Nano-GO has been explored as a nanocarrier for delivery of various drugs including photosensitizers [26]. It has been reported that the photosensitizers loaded onto nano-GO can be more efficiently delivered into cells [27,28]. Since most photosensitizers are highly π-conjugated and, among hydrophobic porphyrin derivatives, the loading of them on the surface of GO can enhance their aqueous solubility and stability. Nano-formulation can further improve the accumulation of photosensitizers to tumors through the enhanced permeability and retention (EPR) effect [29]. However, the highly efficient π–π stacking between the photosensitizers and GO inhibits the ROS generation ability of the photosensitizers [30]. Since GO is highly stable in biological environments [31], release of photosensitizers through passive biodegrading of GO is inefficient. Consequently, the release of the photosensitizers from the surface of GO with a specific design is needed. To the best of our knowledge, such a GO-based nano-formulation of photosensitizers with specific loading and release properties has not yet been explored.

Serum albumin, the most abundant protein in blood plasma, has been approved by FDA for drug delivery based on its high biocompatibility [32]. The encapsulation of various hydrophobic drugs by serum albumin has been extensively studied [33,34]. It has been demonstrated that albumins are highly promising in encapsulation and release of photosensitizers [35]. As most nanoparticles are internalized by cells through a lysosome-mediated channel, the lysosomal cysteine proteases, such as cathepsin B, could induce the intracellular biodegradation of albumin [36,37], triggering the release of encapsulated photosensitizers. Therefore, combination of albumin with other functional materials, such as GO, by the strategy of nanoarchitectonics may be a versatile means for delivery of photoactive drugs towards enhanced PDT [38–40].

Herein, we report a co-assembly strategy for the fabrication of nanohybrids of GO, albumin and photosensitizer for *in vitro* enhanced PDT (Scheme 1). In the nanohybrids, the hybrid system of albumin and GO makes synergistic effect as delivery carriers. The nanohybrids are prepared by loading of the pre-assembled nanoparticles of chlorin e6 (Ce6) and bovine serum albumin (BSA) on the surface of nano-GO. We discover that both the fluorescence and ROS of as-prepared Ce6-BSA-GO nanohybrids in physiological conditions are quenched as compared to free Ce6. Additionally, Ce6–BSA–GO nanohybrids show a much faster internalization and release of Ce6 upon incubation with cells when compared with free Ce6 or a control group of Ce6–GO (Ce6 directly loaded on the surface of GO). *In vitro* PDT results show that Ce6 released from Ce6–BSA–GO nanohybrids recovers its photocytotoxicity upon the cellular uptake. Thus, an enhanced PDT efficacy of Ce6–BSA–GO nanohybrids is demonstrated. Such nanohybrids prepared by encapsulation and loading of photosensitizer by albumin and GO provide a valuable approach to construct smart and biocompatible PDT agents on demand.

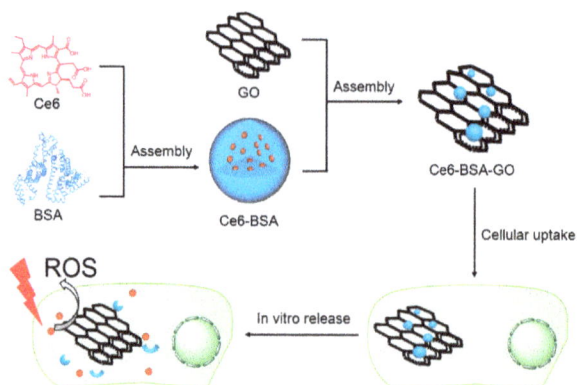

Scheme 1. Schematic illustration of co-assembly of graphene oxide and albumin/photosensitizer nanohybrids and the enhanced *in vitro* photodynamic therapy.

2. Materials and Methods

2.1. Materials

Bovine serum albumin, anthracene-9,10-dipropionic acid (ADPA), thiazolyl blue tetrazolium bromide (MTT), and Hoechst 33342 were purchased from Sigma-Aldrich Co. (Saint Louis, MI, USA) Chlorin e6 was purchased from Frontier Scientific, Inc. (Logan, UT, USA). The cell culture medium DMEM was purchased from Beijing Solarbio Science and Technology Co. Ltd. (Beijing, China). Fetal bovine serum (FBS) was purchased from Hangzhou Sijiqing Co. Ltd. (Hangzhou, China). Other chemicals were purchased from Beijing Chemical Co. Ltd. (Beijing, China) unless otherwise specified. All chemicals were used as received without further purification. Nano graphene oxide was synthesized according to our previous report [25]. Briefly, graphite, $NaNO_3$, and concentrated H_2SO_4 were mixed together in a beaker in an ice bath for 30 min, followed by the slow addition of $KMnO_4$. The reaction mixture was stirred at 35 °C for 6 h, and then the temperature was slowly raised to 60 °C during the next 2 h. The above mixture was then added to water and was stirred at 80 °C for 1 h, followed by adding 30% H_2O_2 and filtering. For purification, the product was alternately washed with 5% of HCl and then water several times. The filter cake was dissolved in water and the graphene oxide flakes were obtained through centrifugation. Finally, the product was freeze-dried in a lyophilizer for two days before use.

2.2. Preparation of the Nanohybrids

The Ce6–BSA–GO nanohybrids were typically prepared as follows: firstly, Ce6-BSA nanoparticles were prepared by mixing a DMSO solution of Ce6 (12 mg·mL^{-1}) with an aqueous solution of BSA (4 mg·mL^{-1}). After aging in the dark for 4 h, the Ce6-BSA nanoparticles were mixed with an aqueous solution of GO (0.5 mg·mL^{-1}). Finally, the obtained nanohybrids were aged overnight and washed by dialysis before further characterization. A control group of Ce6-GO was prepared by mixing a DMSO solution of Ce6 (12 mg·mL^{-1}) with an aqueous solution of GO (0.5 mg·mL^{-1}).

2.3. Characterization of the Nanohybrids

Transmission electron microscopy (TEM) was performed by a JEOL JEM-2100 (Kyoto, Japan) when a drop of the sample was carefully applied to the carbon-coated copper grids and dried in vacuum. A Bruker FastScanBio was used for atomic-force microscopy (AFM) measurements. The size and zeta potential of the nanohybrids was characterized by Malvern dynamic laser scattering (DLS) (Zetasizer Nano ZS ZEN3600, Malvern, UK). UV–Vis spectra of the assembled nanohybrids in aqueous solution were recorded using a Shimadzu UV-2600 spectrophotometer (Kyoto, Japan). A fluorescence spectrometer (Hitachi F-4500, Kyoto, Japan) was used to measure the photoluminescence of free Ce6 and assembled nanohybrids in a 1.0 cm quartz cuvette with the excitation wavelength of 405 nm. The concentrations of Ce6 in samples were measured by a HPLC-based method. Briefly, the sample (100 μL) was mixed with acetonitrile (1 mL) and the obtained mixture was sonicated by an ultrasonic cell crusher (JY92-IIN, Ningbo Scientz Biotechnology Co., Ningbo, China) The concentration of Ce6 in the mixture was further recorded on a Thermo Fisher U3000 HPLC system coupled with a VWD-3100 detector and a reverse phase C18 column (Thermo Scientific Acclaim 120, 5 μm, 4.6 mm × 250 mm, product number 059149, Waltham, MA, USA). Chromatographic conditions: 25 °C; 1.0 mL·min^{-1}; 405 nm; gradient solvent system: v/v acetonitrile/0.1% trifluoroacetic acid in water, and a stepwise gradient of acetonitrile from 45% acetonitrile to 100% in 20 min; t = 9.75 min. In the assembled nanohybrids of Ce6–BSA–GO, the loading amount of Ce6 was evaluated to be 5.5% (w/w).

2.4. ROS Generation

The generation of ROS was detected by the bleaching of 9,10-anthracene dipropionic acid (ADPA). A mixed solution of the nanohybrids and ADPA was prepared. During the following experiment, the solution was stirred vigorously to ensure the saturation of air. When the solution was irradiated by a 635 nm laser (10 mW·cm^{-2}), the bleaching of the absorption band of ADPA at 399 nm was monitored. The solution containing only ADPA was also irradiated by the laser and set as a control group.

2.5. Cell Culture

HeLa cells, generously provided by Prof. Junbai Li (Institute of Chemistry, CAS, Beijing, China), were cultured in the cell culture medium (DMEM supplemented with 10% FBS) at 37 °C under 5% CO$_2$ atmosphere according to standard cell culture protocols.

2.6. In Vitro Imaging

The cells were trypsinized and seeded in a 35 mm glass-bottom dish with a density of 5 × 10^4 cells per well in 2 mL of culture medium. After 24 h, an aliquot of nanohybrids was added to the dish to ensure a Ce6 concentration of 10 μg·mL^{-1}. At pre-determined time points, the cells were washed with phosphate-buffered saline (PBS), stained with Hoechst 33342 fluorescent DNA-binding dye at 5 μg·mL^{-1}, and observed by a confocal laser scanning microscopy (CLSM, Olympus FV1000, Kyoto, Japan). Hoechst 33342 was excited at 405 nm, while Ce6 was excited at 635 nm.

2.7. In Vitro PDT

The cells were seeded to 96-well plates with a density of 2.5 × 10^4 cells per well and were incubated for 24 h. Then, the cells were further incubated with the nanohybrids for 24 h. The cells were then washed with fresh culture medium and irradiated by a 635 nm laser (0.2 W·cm^{-2}) for 1 min. After irradiation, the cells were incubated for another 24 h before the cell viability test by the MTT assay according to the manufacture's protocol.

3. Results and Discussion

3.1. Preparation and Characterization

Ce6 selected for this study is a representative second-generation photosensitizer for PDT [41]. Due to its large molar extinction coefficient in the near-infrared range, Ce6 has been extensively studied for PDT [42]. However, free Ce6 aggregates easily in aqueous solution, disenabling the direct application of free Ce6 for PDT. Ce6–BSA–GO nanohybrids were prepared via two steps, including the preparation of Ce6-BSA nanoparticles and the loading of Ce6–BSA nanoparticles by nano-GO. Ce6–BSA nanoparticles were prepared by co-assembly of Ce6 and BSA. The DLS size distribution shows that Ce6–BSA nanoparticles have a mean hydrodynamic diameter of 38 ± 10 nm (Figure 1a).

TEM image of Ce6–BSA nanoparticles (Figure 1b) indicates that they possess a spherical morphology and their average diameter is almost in agreement with the DLS result. By mixing the pre-assembled Ce6-BSA nanoparticles with nano-GO, negatively-charged Ce6–BSA–GO nanohybrids (zeta potential: -18 ± 5.6 mV) with a mean hydrodynamic diameter of 112 ± 40 nm were obtained. Since nano-GO has a mean hydrodynamic diameter of 53 ± 14 nm, the size increment after mixing suggests the successful loading of Ce6-BSA nanoparticles. The size of graphene oxide is on the scale of 1000 nm, as measured by TEM (Figure 1c), but less than 100 nm in hydrodynamic diameter detected by DLS in solutions. This is because samples are dried in vacuum before characterization by TEM. In such a dry state, GO is in the shape of a film, whereas GO tends to curl in the aqueous solution. That is why the sizes of GO and Ce6–BSA–GO characterized by TEM are much larger than the ones obtained by DLS. TEM images of nano-GO (Figure 1c) and Ce6–BSA–GO (Figure 1d) also confirm that Ce6–BSA nanoparticles are loaded on the surface of nano-GO. AFM image (Figure 1e) and topographic height diagram (Figure 1f) of Ce6–BSA–GO nanohybrids show a height of 10–30 nm, also consistent with the size of Ce6-BSA nanoparticles. The loading of Ce6-BSA on the surface of GO should be the result of hydrophobic and π–π stacking interactions between GO and aromatic residues of BSA as revealed by recent studies [43,44]. Meanwhile, it has been demonstrated that such non-covalent interactions between BSA and GO is strong in a physiological condition of 10 mM phosphate buffer solution at pH 7.4 [43]. Hence, Ce6–BSA–GO nanohybrids are expected to be suitable for drug delivery. For comparison, Ce6–GO nanohybrids with a mean hydrodynamic diameter of 38 ± 10 nm (Figure 1a) were prepared by directly loading of Ce6 on GO.

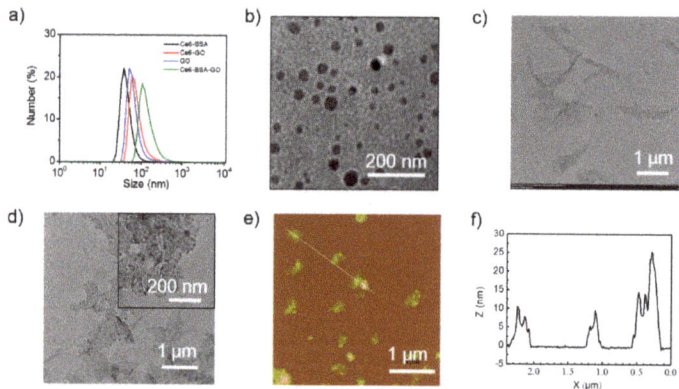

Figure 1. The size and morphology of the nanohybrids. (**a**) Size distribution of Ce6–BSA, Ce6–GO, GO, and Ce6–BSA–GO; (**b**) TEM image of Ce6–BSA; (**c**) TEM image of GO; (**d**) TEM images of Ce6–BSA–GO; (**e**) AFM image of Ce6–BSA–GO; and (**f**) the line profile of Ce6–BSA–GO obtained from (**e**).

UV–Vis absorption spectra of Ce6–BSA–GO nanohybrids, the starting components, and the intermediate materials (Figure 2) were studied to reveal the aggregation status of Ce6 in the nanohybrids. Free Ce6 in water shows a decreased Q-band at 670 nm due to the aggregation induced by intermolecular hydrophobic and π–π stacking interactions [45]. The presence of a characteristic peak at 280 nm of BSA in Ce6–BSA confirms the presence of BSA in the formed Ce6–BSA nanoparticles. Importantly, Ce6–BSA nanoparticles show an increased Q-band, indicating that the aggregation of Ce6 is partly inhibited by the interactions between Ce6 and BSA. This may attribute to the encapsulation of Ce6 in the hydrophobic domains of BSA [46,47]. The absorption bands of Ce6–BSA–GO are similar to those of Ce6–BSA, suggesting the non-covalent interactions between BSA and GO have no significant impact on the absorption of Ce6. By contrast, Ce6–GO shows a significant decreased Q-band, suggesting that a higher degree of aggregation of Ce6 may be induced by the direct contact with GO.

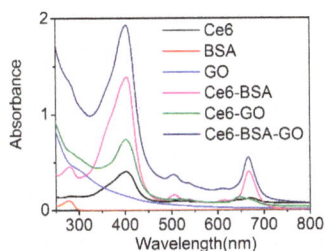

Figure 2. UV–Vis absorption spectra of Ce6, BSA, GO, Ce6-BSA, Ce6-GO, and Ce6–BSA–GO.

3.2. Fluorescence and ROS

In PDT, the generation of cytotoxic ROS is induced by the irradiation. As the light irradiation will be applied to the selected area for a limited time, it is more favorable to inhibit the ROS generation of the photosensitizer before it can be accumulated in the targeted tissue. At present, clinically-applied photosensitizers are always sensitive to light. Hence, special protections are needed in storage of these photosensitizers and the patients must be kept from light for a long time [13]. To examine whether the nanohybrids can inhibit the ROS activity of Ce6, a fluorescence assay was conducted because fluorescence and ROS share the same pathway of excitation and they could be quenched by aggregation simultaneously. In the fluorescence experiment, the emission spectra of Ce6–BSA–GO were compared with various samples containing the same amount of Ce6 (Figure 3). The emission intensities of Ce6–BSA–GO and Ce6–GO are almost identical and are found to be significantly quenched in comparison with those of Ce6 and Ce6–BSA. Although the absorption spectra suggest a low aggregation degree of Ce6 in Ce6–BSA, the emission intensity of Ce6–BSA is found lower than that of Ce6, suggesting that the energy transfer occurs between Ce6 and BSA [4]. In addition to the energy transfer between Ce6 and BSA, an additional energy transfer between Ce6 and GO is also presumably responsible for the low florescence intensity of Ce6–BSA–GO, along with the self-quenching of Ce6.

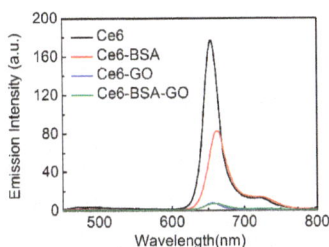

Figure 3. Fluorescence emission spectra of Ce6, Ce6–BSA, Ce6–GO, GO, and Ce6–BSA–GO. The samples were adjusted to contain the same amount of Ce6 and excited at 405 nm.

To monitor the ROS activity of Ce6–BSA–GO directly, ADPA was applied as a ROS-sensitive sensor. ADPA is a water-soluble π-conjugated sensor with a characteristic absorption band at 400 nm [48]. In the presence of ROS, the band at 400 nm decreases due to the breaking of the conjugation. The experiments were conducted by irradiation of the samples containing the same concentration of Ce6 and ADPA. At various time terminals, the absorbance at 400 nm was monitored. When the samples were kept in the dark, no ROS was detected (Figure 4a). When irradiated, Ce6–BSA–GO generated similar amounts of ROS with Ce6–GO, while both of them showed lower ROS activities than Ce6 and Ce6–BSA (Figure 4b). This result is highly consistent with the changes in florescence intensity, indicating the energy transfer between Ce6 and GO inhibits the generation of both fluorescence and ROS. It should be noted that the ROS activity of free Ce6 is partially inhibited due to the self-aggregation.

The further inhibited ROS activity of Ce6–BSA–GO enables a lower sensitivity and a higher stability before the release of Ce6.

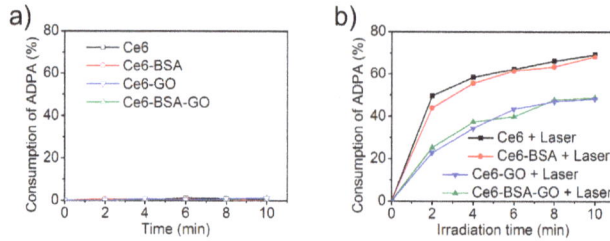

Figure 4. Cumulative consumption of ADPA by the ROS generated by Ce6, Ce6–BSA, Ce6–GO, and Ce6–BSA–GO in the (**a**) absence and (**b**) presence of irradiation.

3.3. Cellular Uptake and In Vitro Release

The ROS, especially singlet oxygen, have a short lifetime and a small diffusion distance in an aqueous environment. As a consequence, the efficient internalization by the cells is crucial for the nanohybrids to achieve a better therapeutic efficiency. In observation of the inhibited ROS activity of Ce6–BSA–GO, we next investigated the cellular internalization and release of Ce6–BSA–GO. CLSM images of the HeLa cells incubated with the nanohybrids for diverse times and stained with Hoechst 33342 show that Ce6–BSA–GO is efficiently internalized by cells and the release of Ce6 from Ce6–BSA–GO is efficiently triggered by cellular microenvironment (Figure 5). Only the fluorescence of Hoechst 33342 is observed in the cells incubated with Ce6 for 24 h. By contrast, the fluorescence of Ce6 is shown in the cytoplasm of the cells incubated with Ce6–GO for 24 h, indicating that the cellular internalization of Ce6–GO is more efficient than free Ce6, which is consistent with a previous report [30].

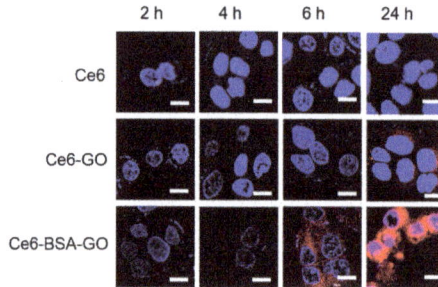

Figure 5. Confocal images of HeLa cells incubated with Ce6, Ce6–GO, or Ce6–BSA–GO for various times and stained by Hoechst 33342. Scale bars denote 15 μm.

For the cells incubated with Ce6–BSA–GO, the fluorescence of Ce6 is obviously shown at 6 h of incubation and significantly enhanced at 24 h of incubation. Significantly, the cells incubated with Ce6–BSA–GO show higher fluorescence intensity than those incubated with Ce6–GO, implying that the Ce6 is more readily released from Ce6–BSA–GO, presumably due to the enzymatic degradation of BSA in lysosomes [49]. The enhanced uptake and release activity of Ce6–BSA–GO allow them to quickly recover the ROS generation ability once accumulated in tumors.

3.4. In Vitro PDT

One advantage of PDT is that the photosensitizers, themselves, are non-toxic to cells in the absence of light. To verify the biocompatibility of the nanohybrids for cells, we incubated HeLa cells with a variety of concentration of nanohybrids in dark for 24 h and checked the cell viabilities by MTT assays. The results show that the cell viabilities are not affected by Ce6, Ce6–GO, or Ce6–BSA–GO at a Ce6 dosage up to 10 μg·mL^{-1} (Figure 6a), indicating the nanohybrids are biocompatible for PDT. To further investigate the photocytotoxicity of the nanohybrids, HeLa cells were incubated with Ce6, Ce6–GO, and Ce6–BSA–GO at a series of Ce6 concentrations for 24 h, followed by irradiation with a 635 nm laser (0.2 W·cm^{-2}) for 1 min. The cell viabilities were determined by MTT assays at 24 h of post-treatment (Figure 6b). The results show that the cell viability was not affected only by light. When samples containing Ce6 were added, the photocytotoxicity started to occur and increased with the increasing Ce6 dosage. Ce6–GO shows a higher photocytotoxicity when compared with Ce6, consistent with the results of *in vitro* fluorescence intensity. The IC50 (the concentration of a photosensitizer inhibits 50% of the cells under light) values of Ce6–BSA–GO and Ce6–GO are 4.5 and 7.8 μg·mL^{-1}, respectively. The lower IC50 of Ce6–BSA–GO than Ce6–GO implies that more efficient release of Ce6 and, consequently, a higher degree of ROS recovery are realized by the application of BSA. The viability of cells pretreated by Ce6–BSA–GO at a Ce6 dosage of 10 μg mL^{-1}, followed by irradiation, decreases to 8.95%. This suggests that a high PDT efficacy is successfully obtained by the nanohybrids co-assembled from photosensitizer, albumin, and GO.

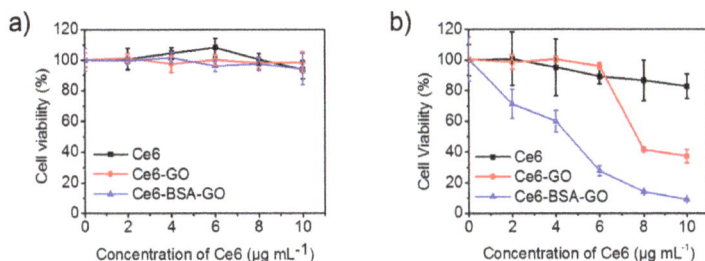

Figure 6. *In vitro* PDT. (**a**) MTT cell viability of HeLa cells incubated with the nanohybrids in dark for 24 h; (**b**) MTT cell viability of HeLa cells incubated with the nanohybrids in dark for 24 h, followed by the irradiation by a 635 nm laser (0.2 W·cm^{-2}) for 1 min.

4. Conclusions

We have demonstrated a smart drug delivery system for photosensitizers by using GO as a ROS quencher and BSA as a biologically-derived degradable component for accelerating the intracellular release. The formed Ce6–BSA–GO nanohybrids show inhibited fluorescence and ROS activities compared to free Ce6 and Ce6-GO. The low ROS activity is beneficial for the protection of a photosensitizer before its accumulation in targeted tissues, such as in storage and in the circulation system. We also demonstrate that the fluorescence and ROS activities of Ce6–BSA–GO are efficiently recovered once they are internalized by cancer cells. The cells incubated with Ce6–BSA–GO are photosensitive and show a more effective PDT that the cells incubated with free Ce6 or Ce6–GO. This proof-of-concept drug delivery system, combining the advantages of binary components of GO and BSA, has the potential to be developed as a robust and versatile tool for delivery of a wide range of photosensitive drugs.

Acknowledgments: We acknowledge financial support from the National Natural Science Foundation of China (Project No. 21522307, 21473208, 91434103 and 51403214), the Talent Fund of the Recruitment Program of Global Youth Experts, the CAS visiting professorships for senior international scientists (Project No. 2016VTA042) and the Chinese Academy of Sciences (CAS). This work was also financially supported by the Natural Science Foundation of China (No. 21473153), the Science Foundation for the Excellent Youth Scholars from Universities and Colleges of Hebei Province (No. YQ2013026), the Support Program for the Top Young Talents of Hebei Province, the Post-graduate's Innovation Fund Project of Hebei Province (No. 2016SJBS009), the China Postdoctoral Science Foundation (No. 2015M580214), and the Scientific and Technological Research and Development Program of Qinhuangdao City (No. 201502A006).

Polymers **2016**, *8*, 181

Author Contributions: Xuehai Yan, Tifeng Jiao and Qianli Zou conceived and designed the experiments; Ruirui Xing, Kai Ma and Yamei Liu performed the experiments; Ruirui Xing, Tifeng Jiao, Xuehai Yan, Qianli Zou and Guanghui Ma analyzed the data; Yamei Liu contributed reagents/materials/analysis tools; Ruirui Xing, Qianli Zou and Xuehai Yan wrote the paper.

Conflicts of Interest: The authors declare no conflict of interest.

Abbreviations

The following abbreviations are used in this manuscript:

PDT	Photodynamic therapy
GO	Graphene oxide
Ce6	Chlorin e6
BSA	Bovine serum albumin
ROS	Reactive oxygen species
EPR	Enhanced permeability and retention
MTT	Thiazolyl blue tetrazolium bromide
FBS	Fetal bovine serum
TEM	Transmission electron microscope
AFM	Atomic-force microscope
DLS	Dynamic laser scattering
ADPA	9,10-anthracene dipropionic acid
PBS	Phosphate buffered saline
CLSM	Confocal laser scanning microscopy

References

1. Castano, A.P.; Mroz, P.; Hamblin, M.R. Photodynamic therapy and anti-tumour immunity. *Nat. Rev. Cancer* **2006**, *6*, 535–545. [CrossRef] [PubMed]

2. Yano, S.; Hirohara, S.; Obata, M.; Hagiya, Y.; Ogura, S.; Ikeda, A.; Kataoka, H.; Tanaka, M.; Joh, T. Current states and future views in photodynamic therapy. *J. Photochem. Photobiol. C* **2011**, *12*, 46–67. [CrossRef]

3. Kachynski, A.V.; Pliss, A.; Kuzmin, A.N.; Ohulchanskyy, T.Y.; Baev, A.; Qu, J.; Prasad, P.N. Photodynamic therapy by *in situ* nonlinear photon conversion. *Nat. Photonics* **2014**, *8*, 455–461. [CrossRef]

4. Plaetzer, K.; Krammer, B.; Berlanda, J.; Berr, F.; Kiesslich, T. Photophysics and photochemistry of photodynamic therapy: fundamental aspects. *Lasers Med. Sci.* **2009**, *24*, 259–268. [CrossRef] [PubMed]

5. Ethirajan, M.; Chen, Y.H.; Joshi, P.; Pandey, R.K. The role of porphyrin chemistry in tumor imaging and photodynamic therapy. *Chem. Soc. Rev.* **2011**, *40*, 340–362. [CrossRef] [PubMed]

6. Qin, C.C.; Fei, J.B.; Wang, A.H.; Yang, Y.; Li, J.B. Rational assembly of a biointerfaced core@shell nanocomplex towards selective and highly efficient synergistic photothermal/photodynamic therapy. *Nanoscale* **2015**, *7*, 20197–20210. [CrossRef] [PubMed]

7. Zou, Q.L.; Zhao, H.Y.; Zhao, Y.X.; Fang, Y.Y.; Chen, D.F.; Ren, J.; Wang, X.P.; Wang, Y.; Gu, Y.; Wu, F.P. Effective two-photon excited photodynamic therapy of xenograft tumors sensitized by water-soluble bis(arylidene)cycloalkanone photosensitizers. *J. Med. Chem.* **2015**, *58*, 7949–7958. [CrossRef] [PubMed]

8. Xing, R.R.; Liu, K.; Jiao, T.F.; Zhang, N.; Ma, K.; Zhang, R.Y.; Zou, Q.L.; Ma, G.H.; Yan, X.H. An injectable self-assembling collagen-gold hybrid hydrogel for combinatorial antitumor photothermal/photodynamic therapy. *Adv. Mater.* **2016**. [CrossRef] [PubMed]

9. Xing, R.R.; Jiao, T.F.; Yan, L.Y.; Ma, G.H.; Liu, L.; Dai, L.R.; Li, J.B.; Mohwald, H.; Yan, X.H. Colloidal gold-collagen protein core-shell nanoconjugate: One-step biomimetic synthesis, layer-by-layer assembled film, and controlled cell growth. *ACS Appl. Mater. Interfaces* **2015**, *7*, 24733–24740. [CrossRef] [PubMed]

10. Chen, Z.A.; Kuthati, Y.; Kankala, R.K.; Chang, Y.C.; Liu, C.L.; Weng, C.F.; Mou, C.Y.; Lee, C.H. Encapsulation of palladium porphyrin photosensitizer in layered metal oxide nanoparticles for photodynamic therapy against skin melanoma. *Sci. Technol. Adv. Mater.* **2015**, *16*, 054205. [CrossRef]

11. Kankala, R.K.; Kuthati, Y.; Liu, C.L.; Lee, C.H. Hierarchical coated metal hydroxide nanoconstructs as potential controlled release carriers of photosensitizer for skin melanoma. *RSC Adv.* **2015**, *5*, 42666–42680. [CrossRef]

12. Orenstein, A.; Kostenich, G.; Roitman, L.; Shechtman, Y.; Kopolovic, Y.; Ehrenberg, B.; Malik, Z. A comparative study of tissue distribution and photodynamic therapy selectivity of chlorin e6, Photofrin II and ALA-induced protoporphyrin IX in a colon carcinoma model. *Br. J. Cancer* **1996**, *73*, 937–944. [CrossRef] [PubMed]

13. Lucky, S.S.; Soo, K.C.; Zhang, Y. Nanoparticles in photodynamic therapy. *Chem. Rev.* **2015**, *115*, 1990–2042. [CrossRef] [PubMed]

14. Zhou, C.; Abbas, M.; Zhang, M.; Zou, Q.L.; Shen, G.Z.; Chen, C.J.; Peng, H.S.; Yan, X.H. One-Step nanoengineering of hydrophobic photosensitive drugs for the photodynamic therapy. *J. Nanosci. Nanotechnol.* **2015**, *15*, 10141–10148. [CrossRef] [PubMed]

15. Zou, Q.L.; Zhang, L.; Yan, X.H.; Wang, A.H.; Ma, G.H.; Li, J.B.; Mohwald, H.; Mann, S. Multifunctional porous microspheres based on peptide-porphyrin hierarchical co-assembly. *Angew. Chem. Int. Ed.* **2014**, *53*, 2366–2370. [CrossRef] [PubMed]

16. Liu, K.; Xing, R.R.; Chen, C.J.; Shen, G.Z.; Yan, L.Y.; Zou, Q.L.; Ma, G.H.; Mohwald, H.; Yan, X.H. Peptide-induced hierarchical long-range order and photocatalytic activity of porphyrin assemblies. *Angew. Chem. Int. Ed.* **2015**, *54*, 500–505. [CrossRef]

17. Yin, P.T.; Shah, S.; Chhowalla, M.; Lee, K.B. Design, Synthesis, and characterization of graphene-nanoparticle hybrid materials for bioapplications. *Chem. Rev.* **2015**, *115*, 2483–2531. [CrossRef] [PubMed]

18. Ji, Q.M.; Honma, I.; Paek, S.M.; Akada, M.; Hill, J.P.; Vinu, A.; Ariga, K. Layer-by-Layer films of graphene and ionic liquids for highly selective gas sensing. *Angew. Chem. Int. Ed.* **2010**, *49*, 9737–9739. [CrossRef] [PubMed]

19. Malik, S.; Vijayaraghavan, A.; Erni, R.; Ariga, K.; Khalakhan, I.; Hill, J.P. High purity graphenes prepared by a chemical intercalation method. *Nanoscale* **2010**, *2*, 2139–2143. [CrossRef] [PubMed]

20. Yoo, J.M.; Kang, J.H.; Hong, B.H. Graphene-based nanomaterials for versatile imaging studies. *Chem. Soc. Rev.* **2015**, *44*, 4835–4852. [CrossRef] [PubMed]

21. Chen, Y.; Tan, C.L.; Zhang, H.; Wang, L.Z. Two-dimensional graphene analogues for biomedical applications. *Chem. Soc. Rev.* **2015**, *44*, 2681–2701. [CrossRef] [PubMed]

22. Dreyer, D.R.; Todd, A.D.; Bielawski, C.W. Harnessing the chemistry of graphene oxide. *Chem. Soc. Rev.* **2014**, *43*, 5288–5301. [CrossRef] [PubMed]

23. Jiao, T.F.; Zhao, H.; Zhou, J.X.; Zhang, Q.R.; Luo, X.N.; Hu, J.; Peng, Q.M.; Yan, X.H. Self-assembly reduced graphene oxide nanosheet hydrogel fabrication by anchorage of chitosan/silver and its potential efficient application toward dye degradation for wastewater treatments. *ACS Sustain. Chem. Eng.* **2015**, *3*, 3130–3139. [CrossRef]

24. Cai, P.; Feng, X.Y.; Fei, J.B.; Li, G.L.; Li, J.; Huang, J.G.; Li, J.B. Co-assembly of photosystem II/reduced graphene oxide multilayered biohybrid films for enhanced photocurrent. *Nanoscale* **2015**, *7*, 10908–10911. [CrossRef] [PubMed]

25. Jiao, T.F.; Liu, Y.Z.; Wu, Y.T.; Zhang, Q.R.; Yan, X.H.; Gao, F.M.; Bauer, A.J. P.; Liu, J.Z.; Zeng, T.Y.; Li, B.B. Facile and scalable preparation of graphene oxide-based magnetic hybrids for fast and highly efficient removal of organic dyes. *Sci. Rep.* **2015**, *5*, 12451. [CrossRef] [PubMed]

26. Goncalves, G.; Vila, M.; Portoles, M.T.; Vallet-Regi, M.; Gracio, J.; Marques, P.A.A.P. Nano-graphene oxide: A potential multifunctional platform for cancer therapy. *Adv. Healthc. Mater.* **2013**, *2*, 1072–1090. [CrossRef] [PubMed]

27. Miao, W.; Shim, G.; Lee, S.; Lee, S.; Choe, Y.S.; Oh, Y.K. Safety and tumor tissue accumulation of pegylated graphene oxide nanosheets for co-delivery of anticancer drug and photosensitizer. *Biomaterials* **2013**, *34*, 3402–3410. [CrossRef] [PubMed]

28. Sahu, A.; Choi, W.I.; Lee, J.H.; Tae, G. Graphene oxide mediated delivery of methylene blue for combined photodynamic and photothermal therapy. *Biomaterials* **2013**, *34*, 6239–6248. [CrossRef] [PubMed]

29. Petros, R.A.; DeSimone, J.M. Strategies in the design of nanoparticles for therapeutic applications. *Nat. Rev. Drug Discov.* **2010**, *9*, 615–627. [CrossRef] [PubMed]

30. Li, F.; Park, S.; Ling, D.; Park, W.; Han, J.Y.; Na, K.; Char, K. Hyaluronic acid-conjugated graphene oxide/photosensitizer nanohybrids for cancer targeted photodynamic therapy. *J. Mater. Chem. B* **2013**, *1*, 1678–1686. [CrossRef]

31. Zhang, X.Y.; Yin, J.L.; Peng, C.; Hu, W.Q.; Zhu, Z.Y.; Li, W.X.; Fan, C.H.; Huang, Q. Distribution and biocompatibility studies of graphene oxide in mice after intravenous administration. *Carbon* **2011**, *49*, 986–995. [CrossRef]

32. Kratz, F. A clinical update of using albumin as a drug vehicle—A commentary. *J. Control. Release* **2014**, *190*, 331–336. [CrossRef] [PubMed]

33. Zhao, F.F.; Shen, G.Z.; Chen, C.J.; Xing, R.R.; Zou, Q.L.; Ma, G.H.; Yan, X.H. Nanoengineering of stimuli-responsive protein-based biomimetic protocells as versatile drug delivery tools. *Chem. Eur. J.* **2014**, *20*, 6880–6887. [CrossRef] [PubMed]

34. Schoonen, L.; van Hest, J.C. Functionalization of protein-based nanocages for drug delivery applications. *Nanoscale* **2014**, *6*, 7124–7141. [CrossRef] [PubMed]

35. Kratz, F. Albumin as a drug carrier: Design of prodrugs, drug conjugates and nanoparticles. *J. Control. Release* **2008**, *132*, 171–183. [CrossRef] [PubMed]

36. Tehle, G.; Sinn, H.; Wunder, A.; Schrenk, H.H.; Stewart, J.C.M.; Hartung, G.; MaierBorst, W.; Heene, D.L. Plasma protein (albumin) catabolism by the tumor itself—Implications for tumor metabolism and the genesis of cachexia. *Crit. Rev. Oncol. Hemat.* **1997**, *26*, 77–100.

37. Langer, K.; Anhorn, M.G.; Steinhauser, I.; Dreis, S.; Celebi, D.; Schrickel, I.; Faust, S.; Vogel, V. Human serum albumin (HSA) nanoparticles: Reproducibility of preparation process and kinetics of enzymatic degradation. *Int. J. Pharm.* **2008**, *347*, 109–117. [CrossRef] [PubMed]

38. Ariga, K.; Li, J.B.; Fei, J.B.; Ji, Q.M.; Hill, J.P. Nanoarchitectonics for dynamic functional materials from atomic-/molecular-level manipulation to macroscopic action. *Adv. Mater.* **2016**, *28*, 1251–1286. [CrossRef] [PubMed]

39. Aono, M.; Ariga, K. The way to nanoarchitectonics and the way of nanoarchitectonics. *Adv. Mater.* **2016**, *28*, 989–992. [CrossRef] [PubMed]

40. Ariga, K.; Ji, Q.M.; Nakanishi, W.; Hill, J.P.; Aono, M. Nanoarchitectonics: A new materials horizon for nanotechnology. *Mater. Horiz.* **2015**, *2*, 406–413. [CrossRef]

41. Celli, J.P.; Spring, B.Q.; Rizvi, I.; Evans, C.L.; Samkoe, K.S.; Verma, S.; Pogue, B.W.; Hasan, T. Imaging and photodynamic therapy: Mechanisms, monitoring, and optimization. *Chem. Rev.* **2010**, *110*, 2795–2838. [CrossRef] [PubMed]

42. Lui, K.; Xing, R.R.; Zou, Q.L.; Ma, G.H.; Mohwald, H.; Yan, X.H. Simple peptide-tuned self-assembly of photosensitizers towards anticancer photodynamic therapy. *Angew. Chem. Int. Ed.* **2016**, *55*, 3036–3039.

43. Kuchlyan, J.; Kundu, N.; Banik, D.; Roy, A.; Sarkar, N. Spectroscopic and fluorescence lifetime imaging microscopy to probe the interaction of bovine serum albumin with graphene oxide. *Langmuir* **2015**, *31*, 13793–13801. [CrossRef] [PubMed]

44. Noh, J.; Son, S.; Kim, Y.; Chae, B.J.; Ku, B.C.; Lee, T.S. Preparation of conjugated polymer dots as a fluorescence turn-on assay for bovine serum albumin by interaction with graphene oxide. *Mol. Cryst. Liq. Cryst.* **2014**, *600*, 170–178. [CrossRef]

45. Adhao, M.; Ahirkar, P.; Kumar, H.; Joshi, R.; Meitei, O.R.; Ghosh, S.K. Surfactant induced aggregation-disaggregation of photodynamic active chlorin e6 and its relevant interaction with DNA alkylating quinone in a biomimic micellar microenvironment. *RSC Adv.* **2015**, *5*, 81449–81460.

46. Chin, W.W.L.; Praveen, T.; Heng, P.W.S.; Olivo, M. Effect of polyvinylpyrrolidone on the interaction of chlorin e6 with plasma proteins and its subcellular localization. *Eur. J. Pharm. Biopharm.* **2010**, *76*, 245–252. [CrossRef] [PubMed]

47. Zou, Q.L.; Liu, K.; Abbas, M.; Yan, X.H. Peptide-modulated self-assembly of chromophores toward biomimetic light-harvesting nanoarchitectonics. *Adv. Mater.* **2016**, *28*, 1031–1043. [CrossRef] [PubMed]

48. Lindig, B.A.; Rodgers, M.A.J.; Schaap, A.P. Determination of the lifetime of singlet oxygen in D_2O using 9,10-anthracenedipropionic acid, a water-soluble probe. *J. Am. Chem. Soc.* **1980**, *102*, 5590–5593. [CrossRef]

49. Bern, M.; Sand, K.M.K.; Nilsen, J.; Sandlie, I.; Andersen, J.T. The role of albumin receptors in regulation of albumin homeostasis: Implications for drug delivery. *J. Control. Release* **2015**, *211*, 144–162. [CrossRef] [PubMed]

polymers

MDPI

Article

The Construction of an Aqueous Two-Phase System to Solve Weak-Aggregation of Gigaporous Poly(Styrene-Divinyl Benzene) Microspheres

Donglai Zhang [1,2], Weiqing Zhou [1], Juan Li [1], Yace Mi [1,2], Zhiguo Su [1] and Guanghui Ma [1,*]

[1]   National Key Laboratory of Biochemical Engineering, Institute of Process Engineering,
      Chinese Academy of Sciences, Beijing 100190, China; dlzhang@ipe.ac.cn (D.Z.); wqzhou@ipe.ac.cn (W.Z.);
      lijuan@ipe.ac.cn (J.L.); miyacetx@163.com (Y.M.); zgsu@ipe.ac.cn (Z.S.)
[2]   University of Chinese Academy of Sciences, Beijing 100049, China
*   Correspondence: ghma@ipe.ac.cn; Tel.: +86-10-8262-7072

Academic Editor: Frank Wiesbrock
Received: 30 January 2016; Accepted: 11 April 2016; Published: 26 April 2016

Abstract: Gigaporous poly(styrene-divinyl benzene) microspheres made via the surfactant reverse micelles swelling method had a controllable pore size of 100–500 nm. These microspheres had unique advantages in biomacromolecule separation and enzymes immobilization. However, the obtained microspheres adhered to each other in the preparation process. Though the weak aggregation could be re-dispersed easily by mechanical force, it will be difficult to scale up. By analyzing the formation mechanism of the aggregates, a method was presented to rebuild the interface between the internal aqueous channel and the external continuous phase by constructing an aqueous two-phase system (ATPS). Based on the ATPS, the method of emulsification, stirring speed, and surfactant concentration in oil phase were optimized. Under the optimum condition (screen emulsification method, 120 rpm for polymerization and 55% surfactant), the microspheres with a controllable particle size of 10–40 μm and a pore size of about 150 nm were obtained. This new method could significantly decrease the weak-aggregation of microspheres.

Keywords: gigaporous microspheres; weak-aggregation; aqueous two-phase system

1. Introduction

Gigaporous microspheres have unique advantages in bioprocess, especially in enzyme immobilization and the purification of biomolecules with large molecular weight and complex structures. Compared to the conventional agarose particles used in bioseparation with pore sizes of 10–30 nm [1,2], gigaporous microspheres have pore sizes up to 100–500 nm [3–6]. When biomolecules such as PEG-protein, viruses, and VLPs (virus-like particles), whose sizes that are 20 nm or larger are purified [7,8], bigger pores lead to good separation, including higher product bioactivity, a faster separation rate, and higher dynamic loading capacity [9–13]. As the carrier materials used for the immobilization of enzymes, the immobilized lipase in gigaporous microspheres also shows much higher specific activity, thermal stability, and storage stability.

At present, multiple methods are developed to prepare gigaporous polymer materials, such as the high internal phase emulsion method [3,14,15], microfluidity [16], the piling of small particles [17], electrospraying [18], and the double emulsification method [19,20]. However, there are some unfavorable characteristics of the products prepared by these methods, such as low specific surface area and mechanical strength. In order to resolve these disadvantages, a new method named the surfactant reverse micelles swelling method has been put forward [4,5]. This method can prepare polymer microspheres with controllable pore sizes of 100–500 nm and favorable mechanical strength; the yield of this method was up to 95% and is feasible for mass production.

The pore-forming mechanism of this method is as follows. A high concentration of surfactant in oil phase assembles into reverse micelles; and, when the oil phase with micelles mixes with the water phase, the reverse micelles in oil droplets absorb water spontaneously and turn into water channels through the droplet. After polymerization, the water channels become through-holes [21]. By adjusting the surfactant concentration, the diameter of pores can be regulated.

However, a new problem appeared in the preparation process. Some particles aggregated weakly. The aggregation can be easily dispersed at about 0.01 MPa pressure, so we did not think this was a big question at first. However, when the pilot test was done, and the production of gigaporous microspheres reached kilogram grade, great effort, such as ultrasonic dispersion or grind, had to be made to re-disperse the aggregate. Could we avoid the occurrence of these aggregations during the preparation process? Common solutions are adding stabilizer in the continuous phase or increasing the rotation speed of emulsion [22,23]. However, in this system, a high concentration of stabilizer polyvinyl alcohol (PVA) (3%) had been added to the water phase, the viscosity of the water phase was very high. If much more of the stabilizer is added to the water, it is hard to clean the products because of the thick water phase [21]. Moreover, the increase in rotation speed would only have a slight influence to the aggregations of microspheres in this system, and also lead to the changing of the particle size.

Thus, a new method to solve this problem based on the formational mechanism of gigapores was put forward in this study. Generally, the stabilizer adsorbs onto the interface of water and oil [21,23,24]. However, in the reverse micelles swelling process, gigapores made the water in them connected to the external continuous aqueous phase; at this region, the interface disappears and the stabilizer cannot adsorb there. During the polymerization process, the particles will impact each other inevitably. Without protection of the stabilizer, the particles will easily adhere to each other.

If the boundary can be rebuilt, the stabilizer can adsorb onto the interface, and the aggregation may be weakened and even disappear [23]. So rebuilding the boundary is a feasible means of reducing the aggregate.

Because the connected region is in the aqueous phase, we were inspired to establish an aqueous two-phase system (ATPS) to rebuild the phase boundary. The ATPS is established by two incompatible water-soluble substances [25–28]. When such substances were dissolved in water, a phase separation would occur. Through this method, the occurrence of the aggregation was effectively weakening, and the gigaporous microspheres with good dispersity were fabricated.

2. Materials and Methods

2.1. Materials

The chemically pure monomer styrene (ST) was purchased from Sinopharm Chemical Reagent Co., Ltd. (Shanghai, China), and the crosslinking agent divinyl benzene (DVB) of commercial grade was obtained from Beijing Chemical Reagents Co., Ltd. (Beijing, China). The Benzoyl peroxide (BPO) with 30% water was used as an initiator and purchased from Sinopharm Chemical Reagent Co., Ltd. (Shanghai, China). Hexadecane (HD) of chromatographic grade was purchased from Haltermann (Houston, TX, USA) and used as a hydrophobic additive. Surfactant sorbitan monooleate (Span 80) of reagent grade was obtained from Farco Chemical supplies (Beijing, China).

Polyvinyl alcohol (PVA), PVA-217, (degree of polymerization 1700, degree of alcoholysis 88.5%), was provided by Kuraray Co., Ltd. (Tokyo, Japan). Polyethylene glycol (PEG) with molecular weights 2000, 6000, 20,000 were purchased from Sinopharm Chemical Reagent Co., Ltd. (Shanghai, China). Dextran T50 (molecular weight about 50,000) were purchased from Seebio Biotech, Inc. (Shanghai, China). NaCl, I_2, BaCl, and ammonium sulfate [$(NH_4)_2SO_4$] were purchased from Xilong Chemical Co., Ltd. (Guangdong, China). Ethanol and acetone were used to precipitate and wash the particles, which purchased from Beijing Chemical Works (Beijing, China).

2.2. Construction of the ATPS Phase Diagram

The common materials to establish the ATPS are two types: polymer–salt and polymer–polymer. In this research, considered the cost and measurement, the materials of the polymer–polymer were limited to dextran/polyethylene glycol (PEG) and PEG/PVA, and the polymer–salt materials are $(NH_4)_2SO_4$/PEG or NaCl/PEG.

If turbidness was observed when these substances were dissolved in water, phase separation would spontaneously occur after a certain time, and an ATPS would be constructed. After the ATPS was formed, the phase diagram could then be delineated. When the two phases separated completely, they were respectively extracted, and the concentration of the two phases were measured to determinate the phase separation point (node) and line (tie-line).

The way to determine the concentration of PEG in the ATPS was via UV spectrophotometry because the PEG had an absorbance in 520 nm with the reaction of I_2 and BaCl, and this absorbance had correlation with the PEG concentration. Therefore, the PEG concentration can be determined by the absorbance–concentration standard curve. When the concentration was determined, the concentration of another component was calculated by the dry weight. After ensuring the node and tie-line, the node was linked with a curve, the binodal of the phase diagram.

2.3. Preparation of Polymeric Microspheres

A recipe used in previous research [4] of gigaporous particle and materials for an ATPS are shown in Table 1: The mixture of ST (monomer), DVB (crosslinking agent), HD (hydrophobic additive), Span 80 (surfactant), and BPO (initiator) were used as the dispersed phase. Water dissolved with PVA (stabilizer) was used as the continuous phase; on this basis, another component such as PEG or dextran was added into the water phase in order to establish an ATPS. According to the phase diagram, the concentration of each phase in the ATPS could be calculated. In this research, a two-step method was used to construct the ATPS between the inner pore and outside (Figure 1). After preparing each of the aqueous phases respectively, the oil phase was dispersed into one of the aqueous phases to form a first emulsion. A certain amount of time was needed to make the micelle swell. Then, the first emulsion was added into another aqueous phase to obtain the ATPS and then started the process of polymerization with the stirring. The characteristics of the emulsion changed with different emulsification and polymerization conditions, but the use of high concentration PVA made a high viscosity and low interface tension of the emulsion.

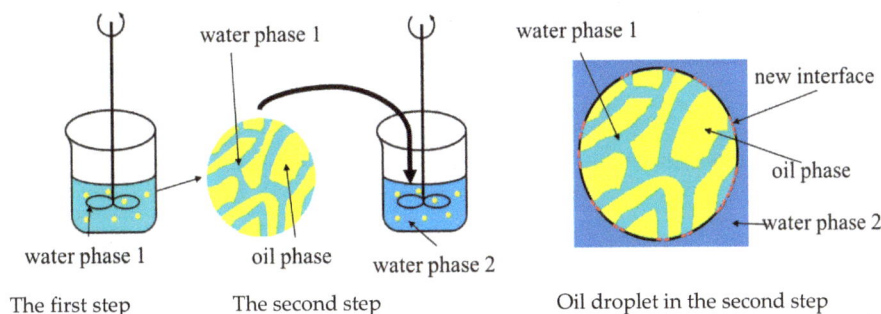

Figure 1. Two-step method to prepare the aqueous two-phase system (ATPS).

Table 1. The recipe for microspheres preparation.

Materials of continuous phase	Weight (g)		Materials of dispersed phase	Weight (g)
	First step	Second step		
PVA	0.2–1.6	0.5–4.0	ST	3.0
Water	17.0–19.5	68.0–78.0	DVB	1.0
PEG(M_w = 2,000, 6,000 and 20,000)	0–0.6	0–3.0	HD	0.4
Dextran	0–0.6	0–3.0	Span 80	2.0–2.4
Ammonium sulfate	0–2.0	0–8.0	BPO	0.8
NaCl	0–2.0	0–8.0		

The polymerization process occurred at 85 °C for 8 h. The polymeric particles were washed with boiled water, ethanol, and acetone three times in turn and dried at 60 °C for 6 h in a glass dish. The yield of particles was calculated by the weight of dried microspheres and the weight of monomers.

2.4. Preparation of Emulsion

In the two-step method, the first step was to make the micelles absorb water and swell, thus determining the size of the oil droplet. In order to get better first emulsion, different methods, including stirring, ultrasonic dispersion, mechanical oscillation membrane emulsification, and screen emulsification methods, were chosen.

For the stirring method, the emulsion was obtained by stirring for 5 to 10 min in 60 to 100 rpm. Additionally, the membrane emulsification method [29,30] was carried out by the use of an SPG membrane, whose pores were from 1 to 50 μm. When the continuous phase and dispersed phase mixed together and were pressed through the membrane by the pressure of the gas, a uniform emulsion could be obtained. Screen emulsification was the same as membrane emulsification, except the SPG membrane was replaced with a large-hole screen.

The second emulsion was prepared by stirring in 120 rpm when the first emulsion was added into another aqueous phase of the ATPS.

2.5. SEM Observation

The aggregation, surface feature, and diameter of the microsphere could be observed by the JSM-6700F scanning electron microscope (SEM) (JEOL, Tokyo, Japan). The dry microspheres in the glass dish were used as a sample; these samples were adhered to the double-sided conductive adhesive tape and then placed on the metal stub. In order to observe, the samples were coated with thin gold film below 5 Pa with a JFC-1600 fine coater.

2.6. Measurement of Particle and Pore Size Distribution

The average diameter of the microsphere and size distribution was measured by laser diffractometry using Mastersizer 2000E (Malvern Instruments Ltd., Malvern, UK).

The pore size distribution was measured by mercury porosimetry measurements—AutoPore IV 9500 mercury porosimetry (Micromeritics, Norcross, GA, USA).

3. Result and Discussion

3.1. Establishment of the ATPS

When the reverse micelles absorbed the water to form a water channel, the phase boundary, which the stabilizer could not absorb on, was lost, and the aggregation occurred. If the water phase inside and outside the channel formed the ATPS, a new boundary would be rebuilt to allow the stabilizer to adsorb, and the aggregation might decrease and even disappear (Figure 2).

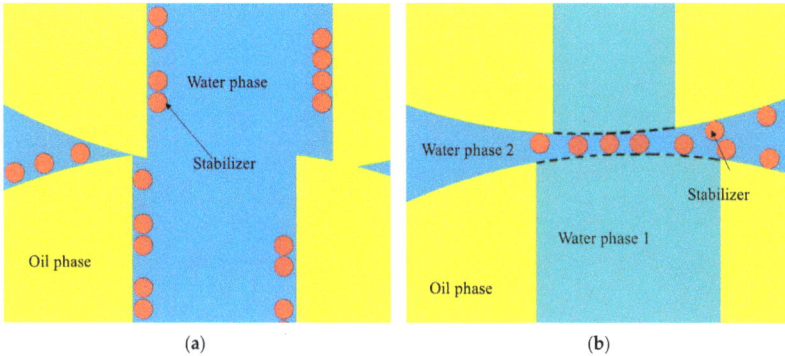

Figure 2. The schematic photographs of boundary loss (**a**) and boundary rebuild (**b**).

Some factors could influence the properties of the ATPS. The materials were the most significant factor. Common materials to establish the ATPS have two types: polymer–salt and polymer–polymer. Comparing the two types, the polymer–salt type usually had a high salt concentration in one phase. High ionic concentration would make the stabilizer coagulation in our experiment. Thus, the polymer–polymer ATPS became a better choice for preparing the gigaporous particles.

For the polymer–polymer ATPS, PEG/dextran or PEG/PVA was chosen. Because PVA is the original component (stabilizer) of the polymerization system, it is more suitable to select PVA than dextran. Another reason is that PVA is much cheaper than dextran when taking into account the cost of production. The molecular weight of PEG was chosen from three different molecular weights (M_w = 2000, 6000 and 20,000). Higher molecular weight led to high viscosity of the water phase; meanwhile, it also formed the ATPS with less concentration. In consideration of the PVA concentration to the reaction, the final ATPS component is PEG 20,000 and PVA 217. These two polymers could form the ATPS with an appropriate concentration and a suitable viscosity.

After choosing the suitable materials, we began to draw the phase diagram. This diagram would provide important guidance in selecting suitable content for the ATPS used in preparing the particle. The temperature for making the ATPS phase diagram is a question that needs to be considered. At room temperature, the swelling step proceed and the boundary lost, and at the polymerization temperature (85 °C) was the time of aggregation. By comparing the phase diagrams made under two different temperatures, we found there was little difference. Therefore, the phase diagram of the PEG/PVA ATPS in 85 °C was chosen for further application (Figure 3). Through this diagram, the concentration and the volume of each phase could be ensured.

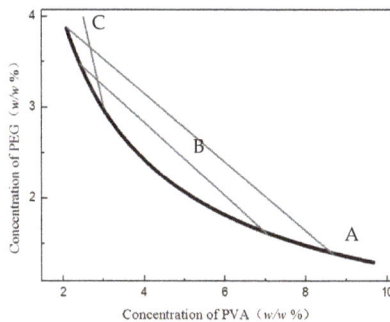

Figure 3. Phase diagram of the polyethylene glycol/polyvinyl alcohol (PEG/PVA) ATPS in 85 °C (A: binodal line; B: tie-lines; C: the concentration ratio chosen in the research).

3.2. Fabrication of Microsphere through the ATPS

In order to rebuild the phase boundary at the droplet, primarily, the interface of the ATPS should close to the initial oil/water interface (Figure 2b), which means that the volume of the aqueous phases inside the oil droplet needs to correspond approximately to the volume of the water channels (the pore volume of the polymerized particles). Furthermore, the volume of the external water phase should be determined by the optimal ratio of the inner phase to the outer phase.

Thus, through calculation, the volume of external water was about 9 times the internal volume. According to the phase diagram, the volume ratio could be determined by the segmentation of tie-line. (Figure 3) In consideration of the stabilization effect and the ease of the after-treatment of the products, the concentration of PVA would be about 3% of the total continuous phase. Therefore, the system concentration of the water phase (C_{total}) and the concentration in the two water phases (C_{inner} and C_{outer}) can be deduced from the figures in Table 2.

Table 2. The concentration of materials in the aqueous two-phase system (ATPS).

Concentration	PEG(*w/w*)	PVA(*w/w*)	PEG/PVA
C_{inner}	1.6%	7.1%	0.23
C_{outer}	3.5%	2.3%	1.52
C_{total}	3.3%	2.8%	1.18

When the concentration of each phase was confirmed, the microspheres could be fabricated. By the two-step method, the phase boundary could be established theoretically. In our experiment, we measured the concentration ratio of the PEG/PVA in the outside water, and the result was nearly 1.5, which was in agreement with the theoretical values of C_{outer} and bigger than C_{total}. This result means that the concentrations of the inner phase and outer phase were not the same (if they were the same, the concentration would be C_{total}), and confirms the formation of the ATPS.

After polymerization, the SEM photo showed that the particle size was one tenth of the microspheres made by the original method (Figure 4a,b), and, for the small microspheres with a diameter of about 10 μm totally aggregated, the aggregation was difficult to disperse so that the size distribution was larger than could be observed. The yield of the reaction was 87% because the small particles were lost in the after-treatment. In addition to the interface effect, another main factor for the aggregation was due to the bigger surface energy of the small particle. Thus, we had to adjust the particle size. Considering that the stirring speed had a significant influence on the particle size, the polymerization was processed with a speed from 150 to 70 rpm. The particle size was from 15 to 25 μm and an aggregated degree was reduced compared with the control group prepared with the same recipe but without the formation of the ATPS (Figures 4c,d and 5). However, we should ascertain whether the existence of the aggregation was due to the small particle or not. Therefore, the next work was aimed at the optimization of the particle size and porous structure of the microspheres in order to find the effect of the ATPS when ruling out the influence of small particles.

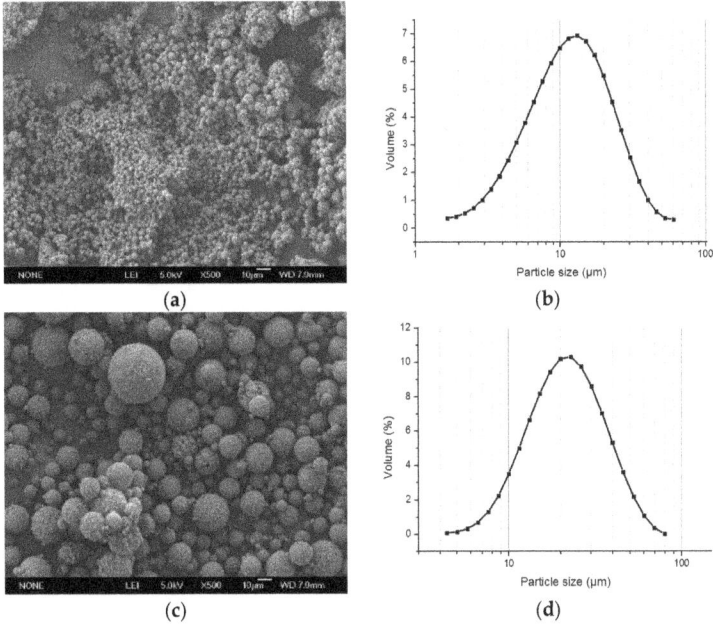

Figure 4. The photographs of the microspheres made at different stirring speeds. Stirring speed 150 rpm, swelling time 10 min (**a**) SEM image (×500) and (**b**) particle size distribution (dispersed). Stirring speed 70 rpm, swelling time 10 min (**c**) SEM image (×500) and (**d**) particle size distribution (dispersed).

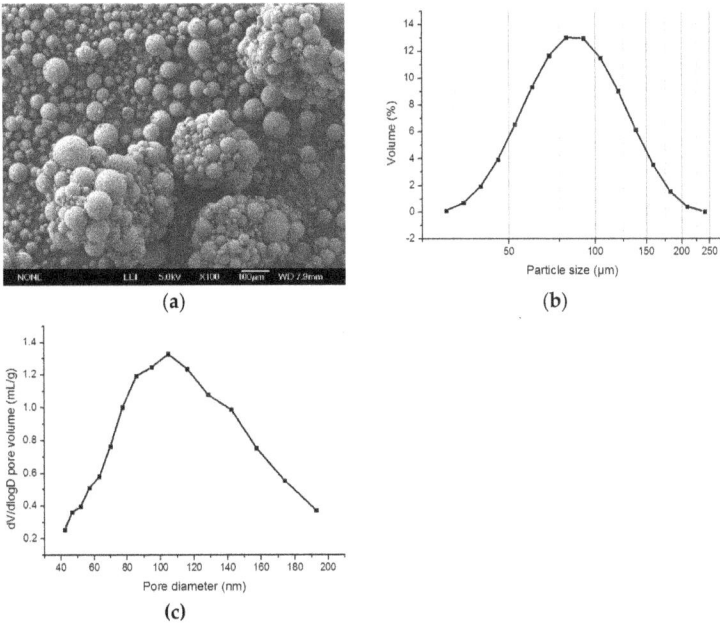

Figure 5. (**a**) The SEM image (×100), (**b**) size distribution (dispersed), and (**c**) pore distribution of control group.

3.3. Optimization of the Particle Size and Porous Structure of the Microspheres

As mentioned above, the significant influence of the stirring rate to the particle size was revealed. Moreover, when the time of the swelling process reduced to half (10 to 5 min), the particle size grew from 25 to 350 μm (Figure 6) and the aggregations were prevented. Therefore, both the stirring speed and swelling time had significantly influenced the particle size (Table 3). Comparing the small droplets with the high surface energy intended to aggregate, the big droplet with lower energy was hard to aggregate. However, the pore size of the particle was only around 100 nm (less than one in a thousand of the particle size), and some of the big particles were hollow. Therefore, emulsion with a proper droplet size was needed.

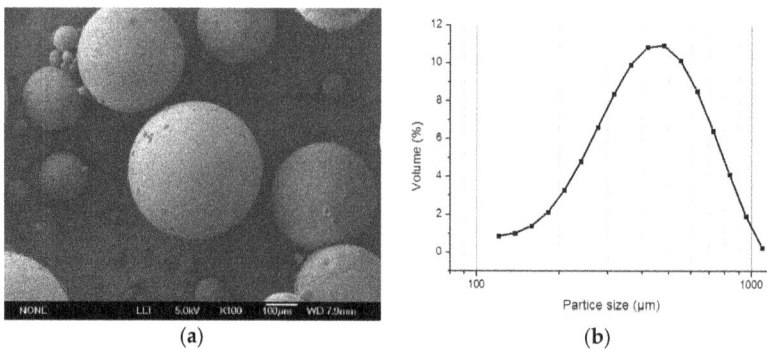

(a)	(b)

Figure 6. (a) The SEM photographs of the microspheres with stirring speed 70 rpm, swelling time 5 min, (×100) and (b) size distribution.

Table 3. The relationship between stirring rate/ swelling time and particle size.

Stirring rate (rpm)	Swelling time (min)	Average particle size (μm)	Uniformity of particle size	Average pore size (nm)	Yield (%)
150	10	10	0.52	80	87
70	10	25	0.33	130	91
70	5	350	0.45	110	93

In order to control the particle size, the membrane emulsification method was used. When using the 50-μm membrane, a large number of small particles formed, almost all of the small microspheres (about 15 μm) aggregated to form larger aggregations (Figure 7), and the pore size of the microspheres was about 100 nm. Considering that the largest pore size of the commercial membrane is 50 μm, a uniform screen with a hole size of around 300 μm was used in the first step of emulsification to increase the size of the oil droplets. After polymerization, the microspheres had a particle size of about 20 μm, and the aggregation was effectively inhibited (Figure 8).

The pore size of the microspheres was about 90 nm (Figure 8c). Because the pore size was positively correlated to the amount of surfactant, we further optimized the content of Span 80 in the oil phase to control the porous structure. The pore characters with a different Span 80 is shown in Table 4. According to the table, when the content of Span 80 was 55% (based on the total amount of ST and DVB), the pore diameter was more than 150 nm (Figure 9, Table 4), which could satisfy the requirement of the separation and purification of large molecules. From the results, it can be concluded that the aggregation was reduced by utilizing the ATPS when excluding the effect of the small particles.

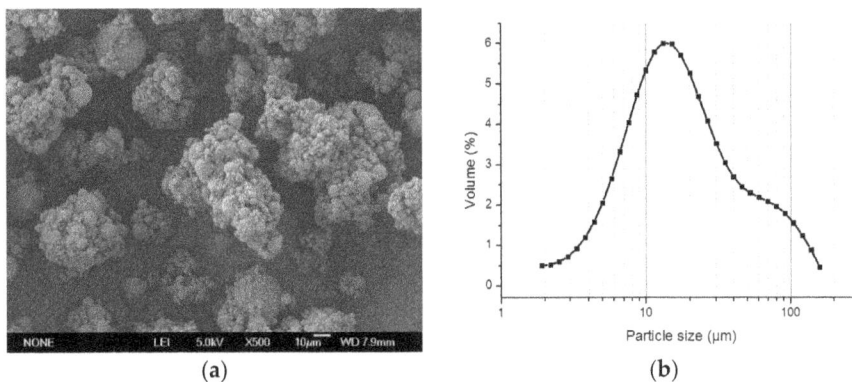

Figure 7. The photographs of the polymer microspheres of the (**a**) membrane emulsification (SEM ×500) and (**b**) size distribution (dispersed).

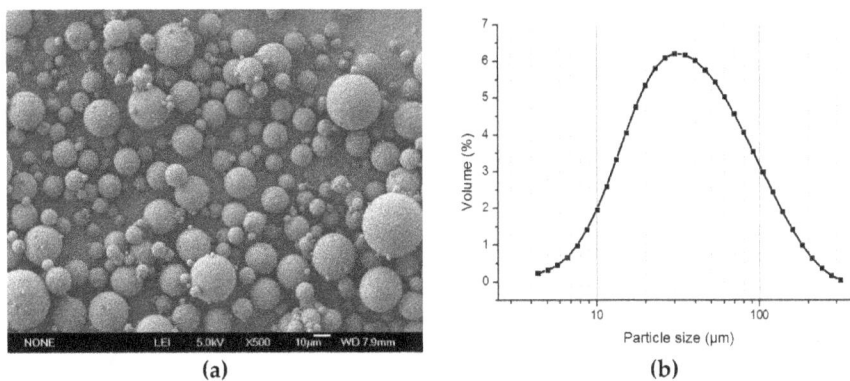

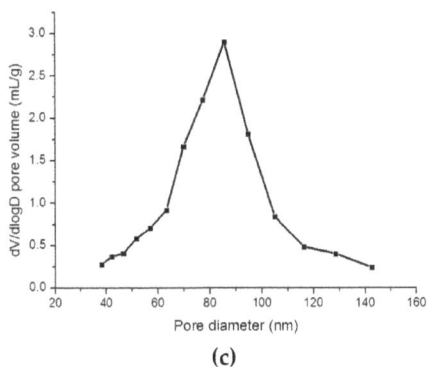

Figure 8. The photographs of the polymer microspheres by screen emulsification (**a**) SEM (×500); (**b**) size distribution; and (**c**) mercury porosimetry measurements.

(a)

(b)

(c)

(d)

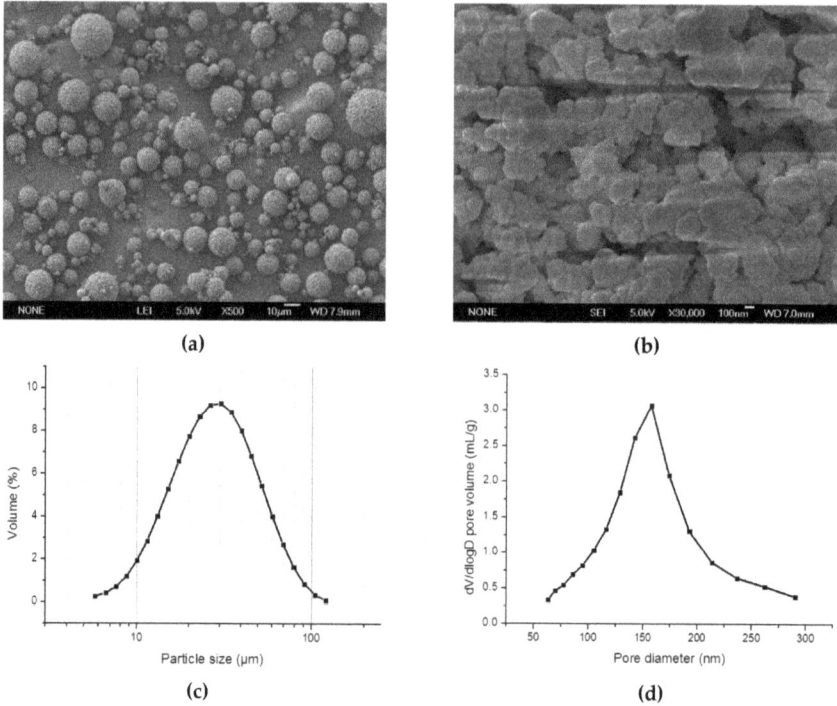

Figure 9. The photographs and mercury porosimetry measurements of the microspheres in optimized concentration of Span 80. (**a**) SEM (×500); (**b**) SEM (×30,000); (**c**) size distribution; and (**d**) mercury porosimetry measurements.

Table 4. The pore characters with different Span 80.

Amount of Span 80 (%) (based on the total amount of ST and DVB)	Average pore size (nm)	Standard error of pore size (nm)	Total pore volume (mL/g)	Porosity (%)	Morphology (watch by SEM)
50	90	22.1	2.06	70	Porous
55	155	40.4	2.87	77	Gigaporous
60	165	48.1	3.60	81	Loose

4. Conclusions

The weak aggregation of gigaporous polymer microspheres occurred when using the reverse micelle swelling method because of the disappearance of the interface between the water channel inside the oil droplet and the outside aqueous phase. A new method, called the aqueous two-phase system (ATPS), was herein proposed to solve the aggregation. A relatively stable boundary between the aqueous channel and the external water environment was established, which provided an interface onto which the stabilizer could adsorb.

The weak aggregation of gigaporous polymer microspheres were effectively reduced by the utilization of the ATPS. Based on the surfactant reverse micelles swelling method and the ATPS, we finally prepared the poly(styrene-divinyl benzene) particles with controllable particle sizes from 10–40 μm and pore sizes from 90–150 nm under the conditions of the screen emulsification method to make first emulsion, 120-rpm stirring for polymerization, and 55% surfactant to produce the pore. This method could reduce the post-treatment of the aggregations, especially for a large amount of products.

Polymers **2016**, *8*, 142

Acknowledgments: The authors acknowledge the financial support by National Natural Science Foundation of China (No. 21336010), the National Science and Technology Support Program (No. 2012AA02A406).

Author Contributions: Guanghui Ma suggested and supervised the work and performed the article editing. Zhiguo Su provided lots of constructive suggestions about this work. Yace Mi provided help in experiments designing and the data analyzing. Juan Li provided the beginning recipe and making process of microspheres. Donglai Zhang and Weiqing Zhou designed and performed the experiments and analyzed the data.

Conflicts of Interest: The authors declare no conflict of interest.

References

1. Barrande, M.; Bouchet, R.; Denoyel, R. Tortuosity of porous particles. *Anal. Chem.* **2007**, *79*, 9115–9121. [CrossRef] [PubMed]
2. Janson, J.C. *Protein Purification: Principles, High Resolution Methods, and Applications*, 3rd ed.; John Wiley & Sons: New York, NY, USA, 2012; p. 532.
3. Yuan, W.; Cai, Y.; Chen, Y.; Hong, X.; Liu, Z. Porous microsphere and its applications. *Int. J. Nanomed.* **2013**, *8*, 1111–1121. [CrossRef] [PubMed]
4. Zhou, W.-Q.; Gu, T.-Y.; Su, Z.-G.; Ma, G.-H. Synthesis of macroporous poly(styrene-divinyl benzene) microspheres by surfactant reverse micelles swelling method. *Polymer* **2007**, *48*, 1981–1988. [CrossRef]
5. Zhou, W.-Q.; Gu, T.-Y.; Su, Z.-G.; Ma, G.-H. Synthesis of macroporous poly(glycidyl methacrylate) microspheres by surfactant reverse micelles swelling method. *Eur. Polym. J.* **2007**, *43*, 4493–4502. [CrossRef]
6. Gokmen, M.T.; Du Prez, F.E. Porous polymer particles—A comprehensive guide to synthesis, characterization, functionalization and applications. *Prog. Polymer Sci.* **2012**, *37*, 365–405. [CrossRef]
7. Zhou, W.; Bi, J.; Janson, J.-C.; Li, Y.; Huang, Y.; Zhang, Y.; Su, Z. Molecular characterization of recombinant Hepatitis B surface antigen from Chinese hamster ovary and *Hansenula polymorpha* cells by high-performance size exclusion chromatography and multi-angle laser light scattering. *J. Chromatogr. B* **2006**, *838*, 71–77. [CrossRef] [PubMed]
8. Mayolo-Deloisa, K.; Lienqueo, M.E.; Andrews, B.; Rito-Palomares, M.; Asenjo, J.A. Hydrophobic interaction chromatography for purification of monoPEGylated RNase A. *J. Chromatogr. B* **2012**, *1242*, 11–16. [CrossRef] [PubMed]
9. Yu, M.; Li, Y.; Zhang, S.; Li, X.; Yang, Y.; Chen, Y.; Ma, G.; Su, Z. Improving stability of virus-like particles by ion-exchange chromatographic supports with large pore size: Advantages of gigaporous media beyond enhanced binding capacity. *J. Chromatogr. A* **2014**, *1331*, 69–79. [CrossRef] [PubMed]
10. Wu, Y.; Abraham, D.; Carta, G. Particle size effects on protein and virus-like particle adsorption on perfusion chromatography media. *J. Chromatogr. A* **2015**, *1375*, 92–100. [CrossRef] [PubMed]
11. Qu, J.-B.; Wan, X.-Z.; Zhai, Y.-Q.; Zhou, W.-Q.; Su, Z.-G.; Ma, G.-H. A novel stationary phase derivatized from hydrophilic gigaporous polystyrene-based microspheres for high-speed protein chromatography. *J. Chromatogr. A* **2009**, *1216*, 6511–6516. [CrossRef] [PubMed]
12. Qu, J.-B.; Zhou, W.-Q.; Wei, W.; Su, Z.-G.; Ma, G.-H. An effective way to hydrophilize gigaporous polystyrene microspheres as rapid chromatographic separation media for proteins. *Langmuir* **2008**, *24*, 13646–13652. [CrossRef] [PubMed]
13. Yao, Y.; Lenhoff, A.M. Pore size distributions of ion exchangers and relation to protein binding capacity. *J. Chromatogr. B* **2006**, *1126*, 107–119. [CrossRef] [PubMed]
14. Hua, Y.; Chu, Y.; Zhang, S.; Zhu, Y.; Chen, J. Macroporous materials from water-in-oil high internal phase emulsion stabilized solely by water-dispersible copolymer particles. *Polymer* **2013**, *54*, 5852–5857. [CrossRef]
15. Wu, R.; Menner, A.; Bismarck, A. Macroporous polymers made from medium internal phase emulsion templates: Effect of emulsion formulation on the pore structure of polyMIPEs. *Polymer* **2013**, *54*, 5511–5517. [CrossRef]
16. Jiang, K.; Sposito, A.; Liu, J.; Raghavan, S.R.; DeVoe, D.L. Microfluidic synthesis of macroporous polymer immunobeads. *Polymer* **2012**, *53*, 5469–5475. [CrossRef]
17. Li, J.; Zhang, Y. Porous polymer films with size-tunable surface pores. *Chem. Mater.* **2007**, *19*, 2581–2584. [CrossRef]

18. Gao, Y.; Bai, Y.; Zhao, D.; Chang, M.-W.; Ahmad, Z.; Li, J.-S. Tuning microparticle porosity during single needle electrospraying synthesis via a non-solvent-based physicochemical approach. *Polymers* **2015**, *7*, 2701–2710. [CrossRef]

19. Chen, S.; Gao, F.; Wang, Q.; Su, Z.; Ma, G. Double emulsion-templated microspheres with flow-through pores at micrometer scale. *Colloid Polym. Sci.* **2012**, *291*, 117–126. [CrossRef]

20. Wang, W.-C.; Peng, C.; Shi, K.; Pan, Y.-X.; Zhang, H.-S.; Ji, X.-L. Double emulsion droplets as microreactors for synthesis of magnetic macroporous polymer beads. *Chin. J. Polym. Sci.* **2014**, *32*, 1639–1645. [CrossRef]

21. Zhou, W.; Li, J.; Wei, W.; Su, Z.; Ma, G. Effect of solubilization of surfactant aggregates on pore structure in gigaporous polymeric particles. *Colloids Surfaces A* **2011**, *384*, 549–554. [CrossRef]

22. Ma, G.; Li, J. Compromise between dominant polymerization mechanisms in preparation of polymer microspheres. *Chem. Eng. Sci.* **2004**, *59*, 1711–1721. [CrossRef]

23. Yang, X.; Liu, L.; Yang, W. Direct preparation of monodisperse core-shell microspheres with surface antibacterial property by using bicationic viologen surfmer. *Polymer* **2012**, *53*, 2190–2196. [CrossRef]

24. Wu, C.; Akashi, M.; Chen, M.-Q. A Simple structural model for the polymer microsphere stabilized by the poly(ethylene oxide) macromonomers grafted on its surface. *Macromolecules* **1997**, *30*, 1–3. [CrossRef]

25. Hatti-Kaul, R.; Kaul, A.; Forciniti, D. *Aqueous Two-Phase Systems*, 1st ed.; Humana Press: New York, NY, USA, 2000; pp. 1–44.

26. Stenekes, R.J.H.; Franssen, O.; Bommel, E.V.M.G.; Crommelin, D.J.A.; Hennink, W.E. The use of aqueous PEG/dextran phase separation for the preparation of dextran microspheres. *Int. J. Pharm.* **1999**, *183*, 1–4. [CrossRef]

27. Ziemecka, I.; van Steijn, V.; Koper, G.J. M.; Rosso, M.; Brizard, A.M.; van Esch, J.H.; Kreutzer, M.T. Monodisperse hydrogel microspheres by forced droplet formation in aqueous two-phase systems. *Lab Chip* **2011**, *11*, 620–624. [CrossRef] [PubMed]

28. Johansson, H.-O.; Karlstrom, G.; Tjerneld, F.; Haynes, C.A. Driving forces for phase separation and partitioning in aqueous two-phase systems. *J. Chromatogr. B* **1998**, 1–15. [CrossRef]

29. Ma, G.-H.; Omi, S.; Nagai, M. Study on preparation and morphology of uniform artificial polystyrene–poly(methyl methacrylate) composite microspheres by employing the SPG (Shirasu Porous Glass) membrane emulsification technique. *J. Colloid Interface Sci.* **1999**, *214*, 264–282. [CrossRef] [PubMed]

30. Ma, G.-H.; Sone, H.; Omi, S. Preparation of uniform-sized polystyrene−polyacrylamide composite microspheres from a W/O/W emulsion by membrane emulsification technique and subsequent suspension polymerization. *Macromolecules* **2004**, *37*, 2954–2964. [CrossRef]

![polymers logo] *polymers*

MDPI

Article

Effect of Small Reaction Locus in Free-Radical Polymerization: Conventional and Reversible-Deactivation Radical Polymerization [†]

Hidetaka Tobita

Department of Materials Science and Engineering, University of Fukui, 3-9-1 Bunkyo, Fukui 910-8507, Japan; tobita@matse.u-fukui.ac.jp; Tel.: +81-776-27-8775

† This is a conference paper from 5th Asian Symposium on Emulsion Polymerization and Functional Polymeric Microspheres (ASEPFPM 2015).

Academic Editor: Haruma Kawaguchi
Received: 29 January 2016; Accepted: 13 April 2016; Published: 20 April 2016

Abstract: When the size of a polymerization locus is smaller than a few hundred nanometers, such as in miniemulsion polymerization, each locus may contain no more than one key-component molecule, and the concentration may become much larger than the corresponding bulk polymerization, leading to a significantly different rate of polymerization. By focusing attention on the component having the lowest concentration within the species involved in the polymerization rate expression, a simple formula can predict the particle diameter below which the polymerization rate changes significantly from the bulk polymerization. The key component in the conventional free-radical polymerization is the active radical and the polymerization rate becomes larger than the corresponding bulk polymerization when the particle size is smaller than the predicted diameter. The key component in reversible-addition-fragmentation chain-transfer (RAFT) polymerization is the intermediate species, and it can be used to predict the particle diameter below which the polymerization rate starts to increase. On the other hand, the key component is the trapping agent in stable-radical-mediated polymerization (SRMP) and atom-transfer radical polymerization (ATRP), and the polymerization rate decreases as the particle size becomes smaller than the predicted diameter.

Keywords: emulsion polymerization; radical polymerization; polymerization rate; theory; reversible-addition-fragmentation chain-transfer (RAFT); stable-radical-mediated polymerization (SRMP); atom-transfer radical polymerization (ATRP)

---

1. Introduction

The rate of polymerization, R_p in free-radical polymerization is represented by:

$$R_p = k_p[M][R^\bullet] \tag{1}$$

where k_p is the propagation rate constant, $[M]$ is the monomer concentration, and $[R^\bullet]$ is the active radical concentration.

At the same monomer concentration, the polymerization rate is higher for larger radical concentration, $[R^\bullet]$. For usual bulk polymerization, $[R^\bullet]$ is determined from the balance of initiation rate R_I and termination rate R_t under steady state, with $R_I = R_t = k_t[R^\bullet]^2$, leading to obtain:

$$[R^\bullet] = \sqrt{\frac{R_I}{k_t}} \tag{2}$$

where k_t is the termination rate constant. Note that the convention of termination rate, $R_t = k_t[R^\bullet]^2$ that does not involve the coefficient 2 is used ([1], p. 12).

The bimolecular termination rate in free-radical polymerization is very fast, and Equation (2) leads to extremely small radical concentration $[R^\bullet]$ in the order of 10^{-8} to 10^{-6} mol/L in usual free-radical polymerization, as schematically represented by (a) bulk polymerization in Figure 1, where red dots are the active radicals.

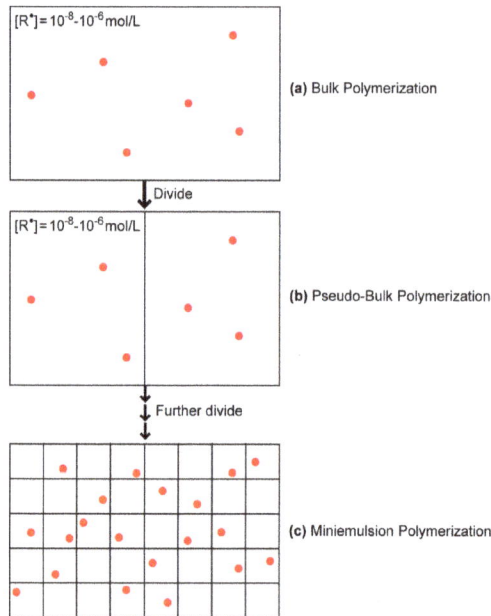

Figure 1. Schematic representation of the radical concentration in (**a**) bulk; (**b**) pseudo-bulk; and (**c**) miniemulsion polymerization.

In general, concentration does not change even when the reaction system is divided, as shown in Figure 1b; pseudo-bulk polymerization. On the other hand, when the polymerization locus is further divided into smaller particles with a few hundred nanometers, the polymerization behavior may change significantly. Each square section in Figure 1c represents a polymer particle in miniemulsion polymerization. The radicals located in other particles cannot terminate each other, and therefore, when the radical generated in the water phase enters a particle without a radical, it may stay to propagate in that particle and the number of radical in it is unity. On the other hand, if the radical enters a particle already having a radical, because the radicals that exist in the same particle possess a very high concentration, they terminate each other instantaneously. Each particle contains zero or one radical, and if the radicals do not exit from the particles, the average number of radicals in a particle is 0.5. This kind of radical isolation effect is sometimes referred to as the compartmentalization ([2], p. 65). The radical concentration in the reaction locus can become much larger than the corresponding bulk polymerization, leading to a much larger rate of polymerization. The average radical concentration in a particle, $[R^\bullet]_p$ is represented by:

$$[R^\bullet]_p = \frac{\bar{n}}{N_A v_p} = \frac{6\bar{n}}{\pi N_A d_p^3} \tag{3}$$

where $\bar{n}$ is the average number of radicals in a particle, N_A is the Avogadro constant, v_p is the particle volume, and d_p is the particle diameter.

Table 1 shows the calculated concentration of a single molecule in a particle with various values of diameter. When the particle diameter is as large as d_p = 1000 nm, the concentration of a single molecule is negligibly small, compared with $[R^\bullet]_{bulk}$, which is in the order of 10^{-8} to 10^{-6} mol/L. The particle with d_p = 1000 nm may contain a large number of radicals in it, and the polymerization behavior is essentially the same as in bulk polymerization. This is the case for suspension polymerization. In general, when the average number of radicals in a particle is larger than 2, the pseudo-bulk polymerization kinetics can be applied [3], at least approximately.

Table 1. Calculated single molecule concentration in a particle.

Particle Diameter d_p (nm)	Concentration (mol/L)
1000	3.18×10^{-9}
200	3.97×10^{-7}
100	3.18×10^{-6}
50	2.55×10^{-5}
25	2.04×10^{-4}

On the other hand, when the particle diameter is smaller than *ca.* 100 nm, even with a single radical in a particle, the concentration may be much larger than usual bulk polymerization, leading to a larger polymerization rate than the corresponding bulk polymerization.

In this article, simple formulas to quantitatively predict the particle diameter below which the polymerization rate becomes much different from the bulk polymerization are elucidated, based on the high single-molecule concentration in a small particle. I already discussed the threshold diameters in the earlier articles [4,5]. However, the statistical variation effect of the key component concentration among the particles was considered at the same time in these articles, which seems to make the discussion rather complicated. In this article, the effect of high single-molecule concentration on the polymerization rate is reorganized, starting from the conventional free-radical polymerization. I mainly refer to the articles of my research group throughout the discussion. This is not because I do not appreciate important contributions by the other research groups deeply, but solely for the theoretical consistency to enhance readability. Readers may find the other interesting aspects of the related topic in refs [2,6–10].

2. Conventional Free-Radical Polymerization

To theoretically consider the effect of small reaction locus, it is convenient to envisage an ideal miniemulsion polymerization. With miniemulson polymerization, the initial stage of polymerization that involves nucleation may be complicated [11]. However, after the initial stage, each polymer particle could be considered as an isolated microreactor, to which a radical enters occasionally. In this article, monodisperse particles are assumed to consider the size effect of polymerization locus. The present discussion applies also for the *ab initio* emulsion polymerization when compared with the corresponding bulk polymerization at the same monomer concentration.

Consider the threshold diameter below which the polymerization rate increases significantly by decreasing the particle size in conventional free-radical polymerization. The average radical concentration in the particles, $[R^\bullet]_p$ is given by Equation (3), which shows that the $[R^\bullet]_p$–value increases significantly by decreasing the particle diameter d_p, assuming the value of $\bar{n}$ does not change notably. When $[R^\bullet]_p$ is larger than that in bulk polymerization, the polymerization rate is larger than the bulk polymerization at the same monomer concentration. Therefore, the condition where the miniemulsion polymerization rate is larger than the corresponding bulk polymerization is represented by:

$$\frac{6\bar{n}}{\pi N_A d_p{}^3} > [R^\bullet]_{bulk} \qquad (4)$$

The radical concentration in bulk polymerization $[R^\bullet]_{bulk}$ is simply given by Equation (2). The particle diameter below which the polymerization rate becomes larger than in bulk polymerization is given by:

$$d^{(1)}_{p,R^\bullet} = \left(\frac{6\bar{n}}{\pi N_A [R^\bullet]_{bulk}} \right)^{1/3} \qquad (5)$$

To represent the threshold diameter, the superscript (1) is used to represent that the effect is caused by a high single-molecule concentration in a small particle, and the subscript $R^\bullet$ represents that the key component is the polymer radical.

Consider a simple example to illustrate the theory. Suppose the initiation rate in bulk polymerization is $R_I = 1 \times 10^{-7}$ mol·L^{-1}·s^{-1}, and the termination rate constant is $k_t = 1 \times 10^7$ L·mol^{-1}·s^{-1}. Assuming a hydrophobic monomer, such as styrene, the average number of radicals for small particles is $\bar{n} = 0.5$ for a wide range of diameters. In this case, Equation (5) leads to the threshold diameter, $d^{(1)}_{p,R} = 252$ nm.

Figure 2 shows the calculated results based on the Monte Carlo (MC) simulation method proposed in [3,12], with $k_p = 500$ L·mol^{-1}·s^{-1}. It is clearly shown that the polymerization rate increases significantly for $d_p < 250$ nm, which shows excellent agreement with Equation (5).

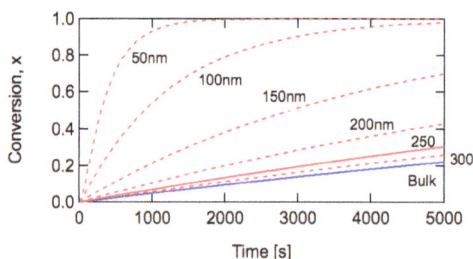

Figure 2. Calculated conversion development for bulk and miniemulsion polymerization with $R_I = 1 \times 10^{-7}$ mol·L^{-1}·s^{-1}, $k_t = 1 \times 10^7$ L·mol^{-1}·s^{-1}, $\bar{n} = 0.5$ and $k_p = 500$ L·mol^{-1}·s^{-1}.

Note that the initiation occurs in the oil phase in bulk polymerization of hydrophobic monomers, and that occurs in water phase in miniemulsion polymerization. Different initiators need to be used and the initiation rate with respect to the unit volume of oil phase is set to be the same in the MC simulation. The initiation rate in water phase R_{Iw} (mol·(L-water)$^{-1}$·s^{-1}) satisfies the following relationship.

$$\frac{R_{Iw}}{N_T} = R_I v_p \qquad (6)$$

where N_T is the total number of particles in unit water phase (L-water)$^{-1}$. The series of miniemulsion polymerizations shown in Figure 2 corresponds to the experiments in which the amount of initiator in the water phase is kept constant and the particle size is changed with a constant monomer/water ratio.

3. Reversible-Deactivation Radical Polymerization

3.1. Polymerization Rate Expression

In free-radical polymerization, the bimolecular termination reactions of the active radicals are inevitable. Therefore, the living polymerization in which the chain termination reactions are totally absent in a strict sense, is impossible. However, if a large percentage of polymer chains are dormant and can potentially grow further, such free-radical polymerization systems can be regarded as pseudo-living polymerization. By introducing the reversible-deactivation process in free-radical polymerization, polymers having a narrow distribution can be obtained. This type of radical polymerization has been referred to as, "controlled", "controlled/living", or "living" radical polymerization. In this article, the IUPAC recommended name [13], reversible-deactivation radical polymerization, RDRP is used.

RDRP belongs to free-radical polymerization, and the polymerization rate expression represented by Equation (1) is still valid. However, to clarify the high single-molecule-concentration effect, it is convenient to use the polymerization rate expression that involves the concentrations of important components and is unique to RDRP.

Figure 3 shows the reversible deactivation reactions in the representative RDRPs, *i.e.*, stable-radical-mediated polymerization (SRMP), atom-transfer radical polymerization (ATRP), and reversible-addition-fragmentation chain-transfer (RAFT) polymerization. In order to formulate the polymerization rate expression for various types of RDRPs in a unified manner, the component that generates an active radical is represented as the radical generating species (RGS), and the component that deactivates an active radical is represented as the trapping agent (Trap) in Figure 3.

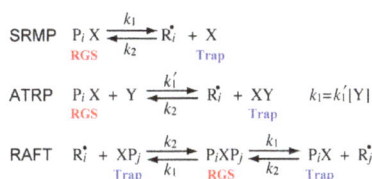

$$
\begin{array}{ll}
\text{SRMP} & \underset{\text{RGS}}{P_i X} \underset{k_2}{\overset{k_1}{\rightleftharpoons}} \underset{}{R_i^\bullet} + \underset{\text{Trap}}{X} \\[2mm]
\text{ATRP} & \underset{\text{RGS}}{P_i X + Y} \underset{k_2}{\overset{k_1'}{\rightleftharpoons}} R_i^\bullet + \underset{\text{Trap}}{XY} \qquad k_1 = k_1'[Y] \\[2mm]
\text{RAFT} & R_i^\bullet + \underset{\text{Trap}}{XP_j} \underset{k_1}{\overset{k_2}{\rightleftharpoons}} \underset{\text{RGS}}{P_i XP_j} \underset{k_2}{\overset{k_1}{\rightleftharpoons}} P_i X + \underset{\text{Trap}}{R_j^\bullet}
\end{array}
$$

Figure 3. Reversible deactivation reaction scheme in each type of RDRP. In the figure, P_iX or XP_i is the dormant polymer with chain length i. R_i is the active polymer radical with chain length i.

Because the lifetime of an active radical in free-radical polymerization is short, normally less than a few seconds, a basic strategy to keep the chain potentially active is to distribute very short active periods throughout the whole reaction time. The rate of deactivation reaction, which is the number of deactivation reactions in a unit volume in a second, is given by:

$$R_{\text{deact}} = k_2[\text{Trap}][R^\bullet] \tag{7}$$

With respect to a single radical, the frequency of deactivation (s^{-1}) is given by:

$$\frac{R_{\text{deact}}}{[R^\bullet]} = k_2[\text{Trap}] \tag{8}$$

Therefore, the average time of a single active period, $\bar{t}_{\text{act}}$ is given by:

$$\bar{t}_{\text{act}} = \frac{1}{k_2[\text{Trap}]} \tag{9}$$

The magnitude of $\bar{t}_{\text{act}}$ is normally in the order of 10^{-4} to 10^{-2} s.

In order to keep a good living condition, the deactivation rate R_{deact} must be much larger than the bimolecular termination reaction rate R_t. If not, a large number of dead polymer chains are formed. At the same time, the activation rate R_{act} must be much larger than the initiation rate R_I. If not, a large number of new chains are formed, leading to not only broadened molecular weight distribution but also to increased termination frequency.

$$R_{\text{deact}} = k_2[\text{Trap}][R^\bullet] \gg R_t \tag{10}$$

$$R_{\text{act}} = k_1[\text{RGS}] \gg R_I \tag{11}$$

For the systems with a very short active period, the polymerization rate is represented by:

$$R_p = R_{gen}\overline{P}_{n,SA} \tag{12}$$

where R_{gen} is the radical generation rate ($R_{gen} = R_{act} + R_I$), and $\overline{P}_{n,SA}$ is the average number of monomeric units added during a single active period.

Because $R_{act} \gg R_I$, the following equation holds.

$$R_{gen} = R_{act} = k_1[RGS] \tag{13}$$

The second term in Equation (12), $\overline{P}_{n,SA}$ can be represented by:

$$\overline{P}_{n,SA} = R_{add}\bar{t}_{act} \tag{14}$$

where R_{add} is the rate of monomer addition to a single active radical, which is given by:

$$R_{add} = \frac{k_p[M][R^\bullet]}{[R^\bullet]} = k_p[M] \tag{15}$$

By substituting Equations (9) and (15) into Equation (14), one obtains:

$$\overline{P}_{n,SA} = \frac{k_p[M]}{k_2[Trap]} \tag{16}$$

From Equations (13) and (16), Equation (12) leads to give the polymerization rate expression unique to RDRP, as follows.

$$R_p = k_p[M]K\frac{[RGS]}{[Trap]} \tag{17}$$

where $K = k_1/k_2$.

Validity of Equation (17) for bulk polymerization under various conditions was examined earlier [4]. The effect of small reaction locus in RDRP is elucidated on the basis of Equation (17).

3.2. SRMP and ATRP

In SRMP and ATRP, the position of equilibrium in the reversible reaction shown in Figure 3 is very much toward the RGS side, and [Trap] << [RGS]. The component whose concentration may become larger by the high single-molecule-concentration effect is the trapping agent. Because [Trap] is in the denominator term in Equation (17), the polymerization rate may become smaller than in bulk polymerization, when the particle size is sufficiently small. The condition where the miniemulsion polymerization rate becomes smaller than in bulk is given by:

$$[Trap]_p > [Trap]_{bulk} \tag{18}$$

where $[Trap]_p$ and $[Trap]_{bulk}$ are the trapping agent concentration in the particle and in bulk polymerization, respectively.

The concentration of a single trapping agent in a particle is given by:

$$(\text{Single trapping agent concentration in a particle}) = \frac{1}{N_A\left(\pi d_p^3/6\right)} \tag{19}$$

Therefore, the miniemulsion polymerization rate is expected to be smaller than in bulk polymerization under conditions represented by the following inequality.

$$\frac{6}{\pi N_A d_p^3} > [\text{Trap}]_{\text{bulk}} \tag{20}$$

The threshold particle diameter, $d_{p,\text{Trap}}^{(1)}$, below which the polymerization rate becomes smaller than the corresponding bulk polymerization is given by:

$$d_{p,\text{Trap}}^{(1)} = \left(\frac{6}{\pi N_A [\text{Trap}]_{\text{bulk}}} \right)^{1/3} \tag{21}$$

The trapping agent concentration in bulk polymerization, $[\text{Trap}]_{\text{bulk}}$ can be determined simply by solving the material balance equation. For SRMP, it can be determined from the following set of differential equations.

$$\frac{d[\text{R}^\bullet]}{dt} = R_I - k_t[\text{R}^\bullet]^2 + k_1[\text{PX}] - k_2[\text{R}^\bullet][\text{X}] \tag{22}$$

$$\frac{d[\text{X}]}{dt} = k_1[\text{PX}] - k_2[\text{R}^\bullet][\text{X}] \tag{23}$$

To examine the validity of Equation (21), the calculated results reported by Zetterlund and Okubo [14] are used. The symbols in Figure 4 show the polymerization rate for the given particle size when the conversion is 10%, reported in [14]. The y-axis shows the ratio of the polymerization rate in miniemulsion and the bulk polymerization $R_p/R_{p,\text{bulk}}$, and the x-axis shows the diameter of particles. The initial trapping agent concentration $[\text{Trap}]_0$ is changed from 0.2 to 0.002 mol/L. For each condition, the threshold diameter $d_{p,\text{Trap}}^{(1)}$ determined from Equation (21) is shown by the red vertical line. The threshold diameter below which the polymerization rate becomes smaller than that in bulk polymerization agrees reasonably well for every condition, which shows that the simple equation, Equation (21) is convenient to estimate the threshold diameter, without conducting complicated calculations.

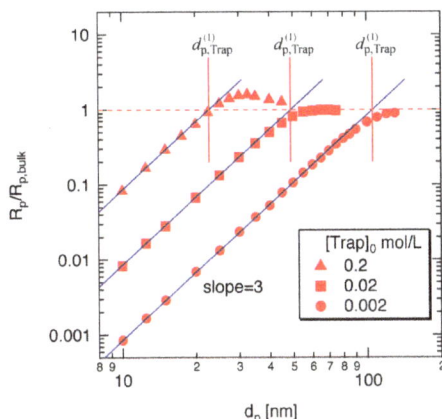

Figure 4. Calculated polymerization rate for the TEMPO-mediated styrene polymerization at 10% conversion with the initial RGS concentration, $[\text{RGS}]_0 = 0.2$ mol/L. The data (symbols) are taken from [14].

Below the threshold diameter, the polymerization rate is proportional to the third power of particle diameter. This is because the single-molecule concentration of [Trap] in a particle, which is in the denominator of Equation (17), is in inverse proportion to the third power of diameter, d_p. Note that [M] and [RGS] inside the particles are large enough to keep the same concentration as the corresponding bulk polymerization.

Figure 4 shows that there exists a particle size region in which the polymerization rate in miniemulsion is slightly larger than in bulk polymerization. This phenomenon results mainly from the statistical variation of the number of trapping agents in a particle [15,16]. In real systems, however, such statistical variation would be blurred by the particle size distribution. In addition, because the degree of acceleration is not very significant, the acceleration region may be difficult to observe experimentally.

In the present theoretical investigation, the exit of trapping agent is not accounted for. If a single trapping agent exits from the polymerizing particle, uncontrolled free-radical polymerization may occur. Because the exit of trapping agent is expected to be more significant for smaller particles, the polymerization rate may not decrease with the third power of d_p. Experimentally, the decrease in polymerization rate by decreasing the particle size is reported [17], but not with d_p^3.

Smaller polymerization rates in smaller polymer particles make it difficult to conduct the *ab initio* emulsion polymerization in SRMP and ATRP.

3.3. RAFT Polymerization

In RAFT polymerization, the concentration of the intermediate, which is RGS in Figure 3, is smaller than that of the trapping agent, and therefore, the high single-molecule concentration effect may be observed for RGS, and [RGS] may become larger than in bulk polymerization. Because [RGS] is the numerator term in the polymerization rate expression given by Equation (17), the polymerization rate may become larger than in bulk, when the particle size is sufficiently small.

Now, consider the RGS concentration in a particle. Practically, a significant increase in polymerization rate, due to the high-single-molecule concentration effect, occurs with the zero-one behavior in conventional free-radical polymerization. The red dashed line in Figure 5 shows the time change of the number of radicals in a particle for the zero-one system. When the second radical enters the particle, two radicals in the particle terminate each other instantaneously. If the exit of a radical can be neglected, the average number of radicals in a particle is $\bar{n} = 0.5$.

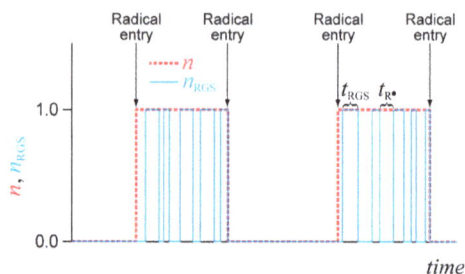

Figure 5. Schematic representation of the zero-one behavior in the conventional free-radical polymerization (**red**) and RAFT (**blue**) miniemulsion polymerization, where n is the number of radicals in a particle in the conventional free-radical polymerization and n_{RGS} is the number of intermediate molecules in RAFT.

When the RAFT agent is introduced, during the growing period in the conventional free-radical polymerization, the number of propagating radicals $n_R{}^\bullet$ and the number of RGS molecules n_{RGS} change zero and one alternatively, because of the reversible reaction shown in Figure 3. The blue line shows the number of RGS molecules in a particle. In RAFT, when the second radical enters the particle, both species must be in the active state in order to cause bimolecular termination reaction. The occurrence of the termination reaction could be delayed slightly compared with the conventional free-radical polymerization. However, the monomer consumption during the delayed period could be neglected. In any case, to roughly estimate the threshold diameter what we want to determine is the approximate value of the average number of RGS molecules in a particle, which could be represented by:

$$(\text{Average number of RGS molecules in a particle}) = \bar{n}\,(1 - \phi_{\text{act}}) \tag{24}$$

where $\bar{n}$ represents the average number of radicals in a particle when no RAFT agent is used, *i.e.*, in the conventional free-radical miniemulsion polymerization, and ϕ_{act} is the average time fraction of the active period, which is defined explicitly by:

$$\phi_{\text{act}} = \frac{\bar{t}_{\text{act}}}{\bar{t}_{\text{act}} + \bar{t}_{\text{inact}}} \tag{25}$$

In the above equation, the average time of a single active period, $\bar{t}_{\text{act}}$ is already given by Equation (9). The average time of a single inactive period, $\bar{t}_{\text{inact}}$ can be formulated similarly with what was done for $\bar{t}_{\text{act}}$, by considering the frequency, $1/\bar{t}_{\text{inact}}$, as follows.

$$\frac{1}{\bar{t}_{\text{inact}}} = \frac{R_{\text{act}}}{[\text{RGS}]} = k_1 \tag{26}$$

Substituting Equations (9) and (26) into Equation (25), one obtains:

$$\phi_{\text{act}} = \frac{K}{K + [\text{Trap}]} \tag{27}$$

The average concentration of RGS in a small particle can be estimated by:

$$[\text{RGS}]_{\text{p}} = \frac{\bar{n}\,(1 - \phi_{\text{act}})}{N_A\,(\pi d_{\text{p}}^3/6)} \tag{28}$$

The miniemulsion polymerization rate is expected to become larger than the corresponding bulk polymerization when $[\text{RGS}]_{\text{p}} > [\text{RGS}]_{\text{bulk}}$. Equation (28) leads to give the following threshold diameter.

$$d_{\text{p,RGS}}^{(1)} = \left(\frac{6\bar{n}\,(1 - \phi_{\text{act}})}{\pi\,N_A[\text{RGS}]_{\text{bulk}}} \right)^{1/3} \tag{29}$$

3.3.1. Two Conflicting RAFT Models

To determine the RGS concentration in bulk polymerization $[\text{RGS}]_{\text{bulk}}$, the material balance equations, similarly with Equations (22) and (23), are needed, which depends on the elementary reactions.

It is known that the RAFT polymerization rate shows retardation behavior by increasing the concentration of the RAFT agent. To rationalize the retardation, two conflicting models were proposed. One model [18] assumes that the intermediate species, PXP, which is an inactive radical, terminates with the propagating radical $R^\bullet$. This model is called the intermediate termination (IT) model. By representing PXP as RGS, the intermediate termination reaction is represented as follows.

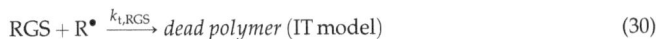

$$\text{RGS} + R^\bullet \xrightarrow{k_{t,\text{RGS}}} dead\ polymer\ (\text{IT model}) \tag{30}$$

where $k_{t,\text{RGS}}$ is the bimolecular termination rate constant between RGS and the active radical, $R^\bullet$.

On the other hand, a slower fragmentation of RGS can also cause retardation [19], which is called the slow fragmentation (SF) model. Both models fit with the bulk polymerization data reasonably well, but the estimated k_1 value for the same reaction system could be more than 10^5 times larger for the IT model than the SF model [20]. The large difference in k_1 leads to a significant difference in the RGS concentration, *i.e.*, $[\text{RGS}]_{\text{bulk,IT}} \ll [\text{RGS}]_{\text{bulk,SF}}$.

The concentration of RGS in bulk polymerization can be determined from the following set of differential equations.

$$\frac{d[\text{R}^\bullet]}{dt} = R_\text{I} + k_1[\text{RGS}] - k_2[\text{R}^\bullet][\text{Trap}] - k_\text{t}[\text{R}^\bullet]^2 - k_{\text{t,RGS}}[\text{R}^\bullet][\text{RGS}] \tag{31}$$

$$\frac{d[\text{RGS}]}{dt} = k_2[\text{R}^\bullet][\text{Trap}] - k_1[\text{RGS}] - k_{\text{t,RGS}}[\text{R}^\bullet][\text{RGS}] \tag{32}$$

$$\frac{d[\text{Trap}]}{dt} = k_1[\text{RGS}] - k_2[\text{R}^\bullet][\text{Trap}] \tag{33}$$

With the SF model, $k_{\text{t,RGS}} = 0$.

The threshold diameter given by Equation (29) shows that a large difference in $[\text{RGS}]_{\text{bulk}}$ leads to a large difference in $d_{\text{p,RGS}}^{(1)}$. Figure 6 shows how $d_{\text{p,RGS}}^{(1)}$ changes during RAFT polymerization, by using a set of representative parameters for dithiobenzoate-mediated styrene polymerization. In the SF model, it takes time to reach the steady state concentration of RGS, and $d_{\text{p,RGS}}^{(1)}$ decreases slowly during the initial stage of polymerization. The threshold diameter is $d_{\text{p,RGS}}^{(1)} = 212$ nm for the IT model, while $d_{\text{p,RGS}}^{(1)}$ is smaller than 10 nm for the SF model.

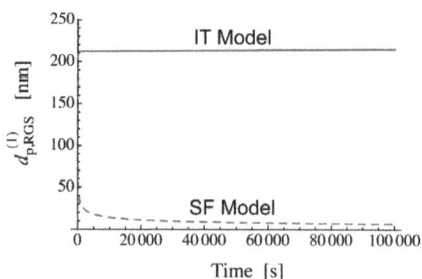

Figure 6. Calculated threshold diameter change during RAFT polymerization. The parameters used are: $R_\text{I} = 1 \times 10^{-7}$ mol·L^{-1}·s^{-1}, $k_\text{p} = 500$ L·mol^{-1}·s^{-1}, $k_\text{t} = 1 \times 10^7$ L·mol^{-1}·s^{-1}, $[\text{M}]_0 = 8$ mol·L^{-1} and $k_2 = 1 \times 10^6$ L·mol^{-1}·s^{-1} for both models. For the IT model, $k_1 = 1 \times 10^4$ s^{-1} and $k_{\text{t,RGS}} = 1 \times 10^7$ L·mol^{-1}·s^{-1}. For the SF model, $k_1 = 0.5$ s^{-1} and $k_{\text{t,RGS}} = 0$.

Figure 7a,b show the simulation results for the conversion development during bulk and miniemulsion polymerization, by using the same set of parameters as in Figure 6. For the miniemulsion polymerization, the MC simulation method proposed earlier [21,22] was used. For the bulk polymerization, the following differential equation for the conversion development was used together with Equations (31)–(33).

$$\frac{dx}{dt} = k_\text{p}(1-x)[\text{R}^\bullet] \tag{34}$$

In the IT model (Figure 7a), the miniemulsion polymerization shows significant rate increase for $d_\text{p} < 212$ nm, as predicted by Equation (29). On the other hand, in the case of the SF model (Figure 7b), the $d_{\text{p,RGS}}^{(1)}$-value is so small (Figure 6), and the polymerization rate is not increased by making the particle size smaller. On the basis of the numerical calculations, using a wide range of parameters, it was concluded that the SF model does not show the polymerization rate increase by decreasing the particle size [23], as long as the given set of parameters cause the retardation behavior in bulk polymerization.

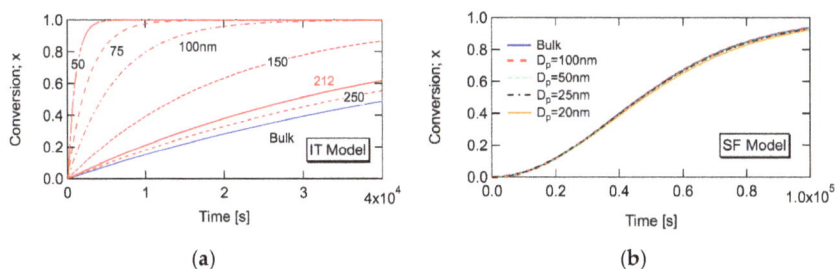

(a)   (b)

Figure 7. Monte Carlo simulation results for IT and SF model, with the same set of parameters used in Figure 6, based on (**a**) the IT model and (**b**) the SF model. Reproduced from Figure 1 in [23]. © Copyright permission from Wiley-VCH Verlag GmbH & Co. KGaA.

3.3.2. Application of Threshold Diameter to Discriminate RAFT Models

On the basis of a large difference in the threshold diameter between the IT model and the SF model, these two models can be discriminated by the miniemulsion polymerization experiment, *i.e.*, a significant polymerization rate increase by decreasing the droplet size is expected for the IT model, while the SF model does not show the rate increase in miniemulsion polymerization. For the miniemulsion experiments, the polymeric RAFT agent is recommended to use to prevent the exit of RAFT agents from the particles.

This model discrimination method was applied for polystyryl dithiobenzoate-mediated styrene polymerization [24,25]. The symbols in Figure 8 show the experimental results. For bulk polymerization, the oil-soluble initiator, AIBN was used, while the water-soluble initiator, potassium persulfate was used in the miniemulsion polymerization experiment. The initiator concentration was adjusted to make the initiation rate per unit volume of oil-phase the same. Good livingness during polymerization was confirmed by the molecular weight distribution development [24,25].

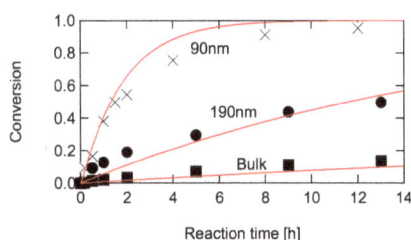

Figure 8. Conversion development during polystyryl dithiobenzoate-mediated styrene polymerization at 60 °C [25]. The solid curves are the calculation results and symbols are the experimental results. For the calculation, the differential equations given by Equations (31)–(34) are solved for the bulk polymerization, and for miniemulsion polymerization, the MC simulation was employed for each fixed diameter. Reproduced from the graphical abstract of [25]. © Copyright permission from Wiley-VCH Verlag GmbH & Co. KGaA.

A significant polymerization rate increase is observed in miniemulsion polymerization by decreasing the particle size, which leads to the conclusion that the IT model applies for the present RAFT polymerization system. The curves are the theoretical calculation results, using the IT model parameters [25]. The conclusion that the IT model applies for the dithiobenzoate-mediate styrene polymerization, rather than the SF model, agrees with the electron paramagnetic resonance (EPR) measurement results [26].

The miniemulsion polymerization is a convenient method for model discrimination. On the other hand, for the accurate estimation of kinetic parameters, bulk polymerization method would be preferable, because it is not disturbed by the existence of water phase as well as the emulsifier.

Larger polymerization rates in smaller polymer particles make it possible to conduct the *ab initio* emulsion polymerization in RAFT systems, by preventing the exit of RAFT agents from the particle. In addition, the RAFT polymerization in sufficiently small reaction loci leads to higher productivity, as in the case of conventional free-radical polymerization.

4. Conclusions

For conventional free-radical polymerization, the threshold particle diameter below which the polymerization rate becomes faster than the corresponding bulk polymerization was derived from the polymerization rate expression, $R_p = k_p[R^\bullet][M]$. On the other hand, for RDRP, the threshold diameter below which the polymerization rate changes significantly compared with the corresponding bulk polymerization was determined from the polymerization rate expression unique to RDRP, $R_p = k_p[M]K[RGS]/[Trap]$. The obtained threshold diameters are summarized in Table 2.

Table 2. Threshold diameter below which the polymerization rate changes significantly compared with the corresponding bulk polymerization.

Type of Polymerization	Threshold Diameter	Polymerization Rate	
Conventional FRP	$d_{p,R^\bullet}^{(1)} = \left(\dfrac{6\bar{n}}{\pi N_A [R^\bullet]_{bulk}} \right)^{1/3}$	Increases for d_p smaller than $d_{p,R^\bullet}^{(1)}$	
SRMP, ATRP	$d_{p,Trap}^{(1)} = \left(\dfrac{6}{\pi N_A [Trap]_{bulk}} \right)^{1/3}$	Decreases for d_p smaller than $d_{p,Trap}^{(1)}$	
RAFT	$d_{p,RGS}^{(1)} = \left(\dfrac{6\bar{n}(1-\phi_{act})}{\pi N_A [RGS]_{bulk}} \right)^{1/3}$	Increases for d_p smaller than $d_{p,RGS}^{(1)}$	

For conventional free-radical polymerization, the polymerization rate increases significantly when the particle size is made smaller than the diameter given by $d_{p,R}^{(1)}$ in the table. Here, the superscript (1) is used to represent that the effect is caused by a high single-molecule concentration in a small particle, and the subscript $R^\bullet$ represents that the key component that causes the polymerization rate change is the polymer radical. The fact that the polymerization rate is faster for smaller particles is one of the reasons why *ab initio* emulsion polymerization is easy to conduct in the conventional free-radical polymerization.

For SRMP and ATRP, the key component to make the polymerization rate slower for smaller polymerization locus is the trapping agent, and the polymerization rate decreases significantly when the particle size is made smaller than the diameter given by $d_{p,Trap}^{(1)}$ in the table.

For RAFT, the key component to make the polymerization rate faster for smaller polymerization locus is the radical generating species (RGS), and the polymerization rate increases significantly when the particle size is made smaller than the diameter given by $d_{p,RGS}^{(1)}$ in the table. This theory can be used to discriminate two controversial models for the RAFT polymerization mechanism, *i.e.*, if the polymerization rate increases significantly in miniemulsion polymerization, the intermediate termination (IT) model applies.

Polymers **2016**, *8*, 155

Acknowledgments: This work is supported by a Grant-in-Aid for Scientific Research, the Ministry of Education, Culture, Sports, Science, and Technology, Japan (Grant-in-Aid 24560938).

Conflicts of Interest: The author declares no conflict of interest.

Abbreviations

The following abbreviations are used in this manuscript:

AIBN	2,2-azobisisobutyronitrile
ATRP	Atom-Transfer Radical Polymerization
IT model	Intermediate Termination model
RAFT	Reversible-Addition-Fragmentation chain-Transfer
RDRP	Reversible-Deactivation Radical Polymerization
RGS	Radical Generating Species
SF model	Slow Fragmentation model
SRMP	Stable-Radical-Mediated Polymerization

References

1. Tobita, H. Polymerization processes, 1. Fundamentals. In *Ullmann's Encyclopedia of Industrial Chemistry*; Wiley-VCH: Weinheim, Germany, 2015.
2. Gilbert, R.G. *Emulsion Polymerization*; Academic Press: London, UK, 1995.
3. Tobita, H.; Takada, Y.; Nomura, M. Simulation model for the molecular weight distribution in emulsion polymerization. *J. Polym. Sci. A* **1995**, *33*, 441–453. [CrossRef]
4. Tobita, H. Modeling controlled/living radical polymerization kinetics: Bulk and miniemulsion. *Macromol. React. Eng.* **2010**, *4*, 643–662. [CrossRef]
5. Tobita, H. Threshold particle diameters in miniemulsion reversible-deactivation radical polymerization. *Polymers* **2011**, *3*, 1944–1971. [CrossRef]
6. Save, M.; Guillaneuf, Y.; Gilbert, R.G. Controlled radical polymerization in aqueous dispersed media. *Aust. J. Chem.* **2006**, *59*, 693. [CrossRef]
7. Oh, J.K. Recent advances in controlled/living radical polymerization in emulsion and dispersion. *J. Polym. Sci. A Polym. Chem.* **2008**, *46*, 6983–7001. [CrossRef]
8. Cunningham, M.F. Controlled/living radical polymerization in aqueous dispersed systems. *Prog. Polym. Sci.* **2008**, *33*, 365–398. [CrossRef]
9. Zetterlund, P.B.; Kagawa, Y.; Okubo, M. Controlled/living radical polymerization in dispersed systems. *Chem. Rev.* **2008**, *108*, 3747–3794. [CrossRef] [PubMed]
10. Zetterlund, P.B. Controlled/living radical polymerization in nanoreactors: Compartmentalization effects. *Polym. Chem.* **2011**, *2*, 534–549. [CrossRef]
11. Bechthold, N.; Landfester, K. Kinetics of miniemulsion polymerization as revealed by Calorimetry. *Macromolecules* **2000**, *33*, 4682–4689. [CrossRef]
12. Tobita, H.; Takada, Y.; Nomura, M. Molecular weight distribution in emulsion polymerization. *Macromolecules* **1994**, *27*, 3804–3811. [CrossRef]
13. Jenkins, A.D.; Jones, R.G.; Moad, G. Terminology for reversible-deactivation radical polymerization previously called "controlled" radical or "living" radical polymerization (IUPAC Recommendations 2010). *Pure Appl. Chem.* **2010**, *82*, 483–491. [CrossRef]
14. Zetterlund, P.B.; Okubo, M. Compartmentalization in TEMPO-mediated radical polymerization in dispersed systems: Effects of macroinitiator concentration. *Macromol. Theory Simul.* **2007**, *16*, 221–226. [CrossRef]
15. Tobita, H. Kinetics of stable free radical mediated polymerization inside submicron particles. *Macromol. Theory Simul.* **2007**, *16*, 810–823. [CrossRef]
16. Tobita, H. Effects of fluctuation and segregation in the rate acceleration of ARTP miniemulsion polymerization. *Macromol. Theory Simul.* **2011**, *20*, 179–190. [CrossRef]
17. Maehata, H.; Buragina, C.; Cunningham, M.; Keoshkerian, B. Compartmentalization in TEMPO-mediated styrene miniemulsion polymerization. *Macromolecules* **2007**, *40*, 7126–7131. [CrossRef]
18. Monteiro, M.J.; Brouwer, H. Intermediate radical termination as the mechanism for retardation in reversible addition-fragmentation chain transfer polymerization. *Macromolecules* **2001**, *34*, 349–352. [CrossRef]

19. Barner-Kowollik, C.; Quinn, J.F.; Uyen Nguyen, T.L.; Heuts, J.P.A.; Davis, T.P. Kinetic investigations of reversible addition fragmentation chain transfer polymerizations: Cumyl phenyldithioacetate mediated homopolymerizations of styrene and methyl methacrylate. *Macromolecules* **2001**, *34*, 7849–7857. [CrossRef]

20. Wang, A.R.; Zhu, S.; Kwak, Y.; Goto, A.; Fukuda, T.; Monteiro, M.S. A difference of six orders of magnitude: A reply to "the magnitude of the fragmentation rate coefficient". *J. Polym. Sci. A* **2003**, *41*, 2833–2839. [CrossRef]

21. Tobita, H.; Yanase, F. Monte carlo simulation of controlled/living radical polymerization in emulsified systems. *Macromol. Theory Simul.* **2007**, *16*, 476–488. [CrossRef]

22. Tobita, H. RAFT miniemulsion polymerization kinetics, 1. Polymerization rate. *Macromol. Theory Simul.* **2009**, *18*, 108–119. [CrossRef]

23. Tobita, H. On the discrimination of RAFT models using miniemulsion polymerization. *Macromol. Theory Simul.* **2013**, *22*, 399–409. [CrossRef]

24. Suzuki, K.; Nishimura, Y.; Kanematsu, Y.; Masuda, Y.; Satoh, S.; Tobita, H. Experimental validation of intermediate tertmination in RAFT polymerization with dithiobenzoate via comparison of miniemulsion and bulk polymerization rate. *Macromol. React. Eng.* **2012**, *6*, 17–23. [CrossRef]

25. Suzuki, K.; Kanematsu, Y.; Miura, T.; Minami, M.; Satoh, S.; Tobita, H. Experimental method to discriminate RAFT models between intermediate termination and slow fragmentation via comparison of rates of miniemulsion and bulk polymerization. *Macromol. Theory Simul.* **2014**, *23*, 136–146. [CrossRef]

26. Meiser, W.; Buback, M.; Ries, O.; Ducho, C.; Sidoruk, A. EPR study into cross-termination and fragmentation with the phenylethyl-phenylethyl dithiobenzoate RAFT Model System. *Macromol. Chem. Phys.* **2013**, *214*, 924–933. [CrossRef]

polymers

MDPI

Article

Towards A Deeper Understanding of the Interfacial Adsorption of Enzyme Molecules in Gigaporous Polymeric Microspheres

Weichen Wang [1,2], Weiqing Zhou [1], Wei Wei [1], Juan Li [1], Dongxia Hao [1], Zhiguo Su [1] and Guanghui Ma [1,*]

[1] National Key Laboratory of Biochemical Engineering, Institute of Process Engineering,
Chinese Academy of Sciences, Beijing 100190, China; wchwang@ipe.ac.cn (W.Wa.); wqzhou@ipe.ac.cn (W.Z.);
weiwei@ipe.ac.cn (W.We.); lijuan@ipe.ac.cn (J.L.); dxhao@ipe.ac.cn (D.H.); zgsu@ipe.ac.cn (Z.S.)
[2] University of Chinese Academy of Sciences, Beijing 100049, China
* Correspondence: ghma@ipe.ac.cn; Tel.: +86-10-8262-7072

Academic Editors: Katja Loos and Frank Wiesbrock
Received: 26 January 2016; Accepted: 24 March 2016; Published: 7 April 2016

Abstract: Compared with the one immobilized in the conventional mesoporous microspheres, the enzyme immobilized in gigaporous microspheres showed much higher activity and better stability. To gain a deeper understanding, we herein selected lipase as a prototype to comparatively analyze the adsorption behavior of lipase at interfaces in gigaporous and mesoporous polystyrene microspheres at very low lipase concentration, and further compared with the adsorption on a completely flat surface (a chip). Owing to the limited space of narrow pores, lipase molecules were inclined to be adsorbed as a monolayer in mesoporous microspheres. During this process, the interaction between lipase molecules and the interface was stronger, which could result in the structural change of lipase molecular and compromised specific activity. In addition to monolayer adsorption, more multilayer adsorption of enzyme molecules also occurred in gigaporous microspheres. Besides the adsorption state, the pore curvature also affected the lipase adsorption. Due to the multilayer adsorption, the excellent mass transfer properties for the substrate and the product in the large pores, and the small pore curvature, lipase immobilized in gigaporous microspheres showed better behaviors.

Keywords: lipase; adsorption; gigaporous; mesoporous; microsphere; QCM-D

1. Introduction

Immobilization is one of the most effective approaches to improve the catalytic performance of enzymes. To gain a better performance, the immobilized carriers should be rationally designed and selected. It has been demonstrated that the physical properties of carriers, such as surface morphology [1,2], particle size [3,4], and inner porous structures [5–8] have an important impact on enzyme immobilization. Recently, many porous particles have also exhibited their distinctive performance as immobilized carriers [9,10]. It is worth noting that the pore size of these conventional carriers usually ranges from a few nanometers to tens of nanometers [11,12]. In this case, the enzyme penetration in most studies might be limited by the tiny pore size. The effect of pore size on the activity of enzymes has been a controversial issue. Because of the complexity of enzymes, different or even opposite results have been reported [13–16].

In our previous work, lipase was immobilized in polystyrene (PST) microspheres with different pore sizes of 14 nm (mesoporous), 100 nm (macroporous), and 300 nm (gigaporous). Fortunately, we found the specific activity of lipase immobilized in gigaporous microspheres was 2.87 times and 1.46 times more than that of mesoporous ones and the free lipase, respectively. In addition, the reusability and the stability of lipase in gigaporous microspheres was improved dramatically compared with those of mesoporous ones [17]. The significance of pore size has been demonstrated in this comparative investigation. Therefore, the reasons of these results and the detailed analysis of enzyme adsorption needed to be researched. There were many reports about the enzyme adsorbed in macroporous microspheres. Gross *et al.* studied the distribution and the molecular structure of lipase adsorbed in microspheres with different pore sizes. The average pore sizes of the carriers were usually 40–100 nm. The results showed that the enzyme is localized in an external shell of the bead [18–21]. They noticed the influences of pore size on the diffusion—the higher the pores, the better the activity due to the diffusion—however, they did not specialize the adsorption state. In addition, although macroporous microspheres were frequently used in enzyme adsorption, many studies mainly focused on the microspheres with a maximum pore size of 100 nm, which may be restricted by the preparation technology of microspheres. As for other researches about gigaporous microspheres, Miletić *et al.* investigated the poly(GMA-*co*-EGDMA) [poly(glycidyl methacrylate-*co*-ethylene glycol dimethacrylate)] microspheres with large pore size (over 300 nm) as the immobilized carriers of lipase. The main conclusions mostly focused on the immobilization and catalysis results [22], yet the effects of the adsorption state were not taken into consideration.

In this work, lipase B from *Candida antarctica* was selected as the prototype because of its wide applications. The adsorption time and the initial concentration of lipase were tailored to investigate the adsorption amount and the enzyme activity of lipase in the microspheres more precisely. Moreover, considering the possible effect of the pore curvature, lipase adsorption in microspheres was compared with that on the flat surface. In order to attain the real-time data of lipase adsorbed on flat surface, quartz crystal microbalance with dissipation monitoring (QCM-D) was used to monitor the adsorption process of enzyme molecules on flat surface [23].

The work carried out on the behaviors of lipase adsorption in the gigapores, in the mesopores, and on the flat surface, which was expected to provide a model for the enzyme immobilization in gigaporous microspheres.

2. Materials and Methods

2.1. Materials

Lipase B from *Candida antarctica* (CALB) (3.8 U/mg by activity assay) was kindly provided by Novozymes (Bagsværd, Denmark). The substrate *p*-nitrophenyl palmitate (*p*-NPP) of analytical grade was purchased from Sigma-Aldrich (St. Louis, MO, USA). The standard *p*-nitrophenol (*p*-NP) of analytical grade was purchased from Sinopharm Chemical Reagent Beijing Co., Ltd. (Beijing, China). The PST chips were purchased from Biolin Scientific AB (Goteborg, Sweden). The giga-/meso-porous PST microspheres were prepared by National Engineering Research Center of Biotechnology (Beijing, China). All chemicals were of analytical grade and used without further purification unless otherwise described.

The scanning electron microscope (SEM, JEOL, JSM-6700F, Tokyo, Japan) images of the microspheres were shown as Figure 1. The structural data of the microspheres measured by mercury porosimetry (Micromeritics, AutoPore IV 9500, Norcross, GA, USA) were shown in Table 1. The gigaporous PST microspheres were recorded as PST-300, and the mesoporous ones were PST-14.

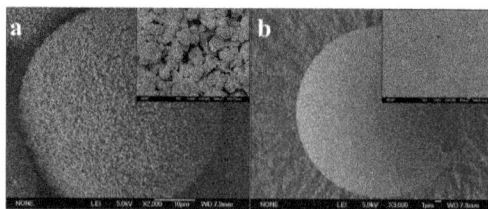

Figure 1. SEM images of PST-300 (**a**) and PST-14 (**b**) microspheres.

Table 1. Structural data of PST-300 and PST-14 microspheres.

Microspheres	Average Pore Size (nm)	Total Pore Surface Area (m²/g)	Total Pore Volume (cm³/g)	Porosity (%)
PST-300	340.3	14.7	1.9	70.4
PST-14	14.5	738.4	2.7	78.7

2.2. Adsorption of Lipase in PST Microspheres

Several groups of giga-/meso-porous PST microspheres (0.01 g) were added into the lipase solution (0.05 mg/mL, 2 mL). Then, the mixture was shaken gently at 25 °C. The adsorption time was 1, 2, 5, 8, 10, 20, 30 min, respectively. When the adsorption time was up to the pre-set value, the corresponding sample was taken out and centrifuged three times to separate the microspheres and the solutions (10,000 rpm, 3 min), and the lipase that was not adsorbed was washed out.

2.3. Assay of Adsorption Mass and Activity of Lipase in PST Microspheres

The amount of lipase adsorbed in microspheres was measured via the bicinchoninic acid (BCA) method at 562 nm using multimode microplate readers (Tecan, Infinite 200, Zurich, Switzerland). The protein concentration of initial lipase solution and residual supernatant were analyzed to determine the adsorption mass. The standard work agent was prepared by the mixture of BCA agent and Cu agent (50:1, *v/v*). A calibration curve was constructed from BSA solutions of known concentration (31.25–2000 µg/mL) and used to calculate the protein amount in initial and washing solutions [24].

The activity of lipase adsorbed in microspheres was also assayed by *p*-NPP hydrolysis. *p*-NPP can be hydrolyzed by lipase to *p*-NP. Firstly, the *p*-NP standard solution was prepared, and the calibration curve was the result. *p*-NPP was dissolved in acetone and then diluted with phosphate buffer (50 mmol/L, pH 7.0) containing Triton X-100 (1.25%, *w/v*). The *p*-NPP and immobilized lipase were mixed in a centrifuge tube with cap for 10 min at 37 °C. Then, the absorbance was monitored at 410 nm using multimode microplate readers.

2.4. CLSM Analysis of Lipase Adsorbed in Microspheres

Confocal laser scanning microscopy (CLSM) (Leica, TCS SP5, Wetzlar, Germany) was employed to investigate the distribution of lipase adsorbed in the microspheres at the time point of 2, 8, 30, 40 min. Fluorescamine was used as a reagent for the detection of lipase, which was soluble in acetone at 50 mg/mL. The fluorescamine solution was added to react with PST-lipase for 3 min. The samples were excited at 390 nm, and the fluorescent images at 480 nm wavelengths were then obtained.

2.5. QCM-D Measurement

Phosphate buffer was filtered through membrane of 0.22 µm and then ultrasonic degassed for 20 min. The QCM-D (Q-SENSE E4, Biolin Scientific AB, Goteborg, Sweden) measurement was conducted at 37 °C, and PST chip was cast into the measure chamber. Phosphate buffer was injected into the chamber. When the baseline was stabilized, a diluted lipase solution was injected into the chamber using a flow rate of 50 µL/min. This process simulated the immobilization procedure. Adsorption time of lipase on PST chip was 90 min from the optimized results of pre-experiments. Then, the chip was rinsed by the phosphate buffer, and the desorbed lipase molecules were rinsed out. When the baseline was stabilized again, *p*-NPP was injected into the chamber to simulate the hydrolysis reaction of *p*-NPP catalyzed by the immobilized lipase. The *p*-NPP mixture (0.05 mg/mL) comprised *p*-NPP (0.05 mg), acetone (12.5 µL), and phosphate buffer (987.5 µL, 0.01 mol/L, pH 7.2).

The determined parameters were Δf and ΔD. Δf represented the frequency change on the chip surface, which reflected the change of adsorption mass. ΔD represented the change of dissipation factors, which reflected the change of adsorption viscoelastic. The $|\Delta D/\Delta f|$ ratio reflected the adsorption tightness, which was contrariwise proportional to the adsorption tightness. A large $|\Delta D/\Delta f|$ ratio indicated an extended structure or loose binding between the interacting molecules [25].

At the same time, the adsorption and the catalysis process was monitored, as a function over time, by simultaneously recording the shifts in the frequency (Δf) and in the energy dissipation (ΔD) at the fundamental resonant frequency, along with the third, fifth, and seventh overtones, until the adsorption reached a steady-state. At this time, the long-term stability of the frequency was within 1 Hz, which was negligible when compared with the frequency shifts caused by adsorption. Normalized data obtained from different overtones were used in the calculation of adsorption mass, thickness, and viscoelastic properties of adsorbed layers using the Voigt model. Sauerbrey mass was calculated using the Sauerbrey equation [26]:

$$M = - (C/n) \Delta f \tag{1}$$

where Δf, M, and n represented frequency change, adsorbed mass per unit area, and overtone number, respectively. C was the mass sensitivity constant (17.7 ng/cm^2 Hz). The adsorption mass was inversely proportional to the Δf value.

2.6. Assay of Adsorption Mass and Activity of Lipase on PST Chip

The adsorption mass of lipase on PST chip can be calculated by Equation (1), of which the unit was ng/cm^2. The activity can be calculated by the decreased amount of the substrate p-NPP. The activity unit (U) of lipase was defined as the amount of lipase required to hydrolyze 1 μmol of p-NPP per minute at 37 °C (pH 7.0). Then, the lipase activity was calculated by:

$$u = \frac{(M_1 - M_2) \times A_{chip} \times 10^6}{377.52 \times t \times M_L} \tag{2}$$

where M_1 was the initial adsorption mass of p-NPP per unit area, M_2 was the adsorption mass of p-NPP per unit area after hydrolysis, A_{chip} was the area of PST chip, of which the value was 1.54 cm^2, the value 377.52 was the molecular weight of p-NPP, M_L was the lipase dosage, and t was the reaction time.

In addition, the specific activity of lipase was defined as the activity of an enzyme per milligram of total protein (U/mg protein).

3. Results and Discussion

3.1. Analysis of Lipase Adsorption in PST Microspheres

Lipase immobilized in the gigaporous PST microspheres showed higher activity, stability, reusability, and better kinetic performance than that in the mesoporous PST ones. In order to attain the detailed adsorption behaviors of lipase, PST-300 and PST-14 were used for the real-time adsorption of lipase. Figure 2 showed the adsorption process of lipase in PST-300 and PST-14, including the change of adsorption mass and lipase activity within 120 min. It can be seen from Figure 2a that the adsorption reached equilibrium for lipase in PST-300 after the time point of 30 min and for lipase in PST-14 after the time point of 60 min. From Figure 2b, it can be seen that the initial phase of lipase adsorption (before the time point of 30 min) had great influences on lipase activity.

In order to study the adsorption of lipase more elaborately, the experiments were carried out firstly under the condition of low lipase concentration (0.05 mg/mL) and short adsorption time (within 30 min). The adsorption states of lipase in gigaporous and mesoporous microspheres were analyzed comparatively. The changes of adsorption mass, enzyme activity, and specific activity of lipase in PST microspheres with adsorption time are shown as Figure 3, which shows a clear difference between the two types of microspheres.

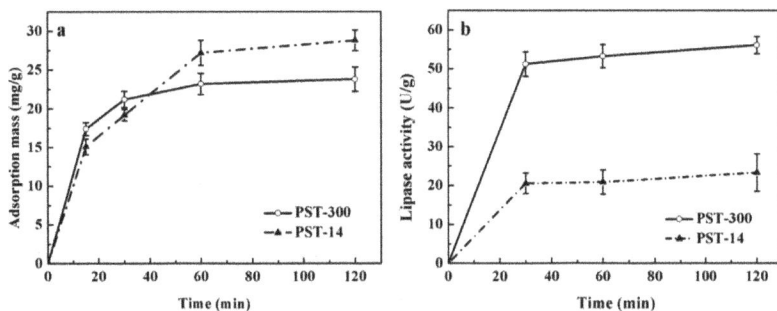

Figure 2. Adsorption process of lipase (0.05 mg/mL) in PST microspheres within 120 min ((a) Adsorption mass; (b) Lipase activity).

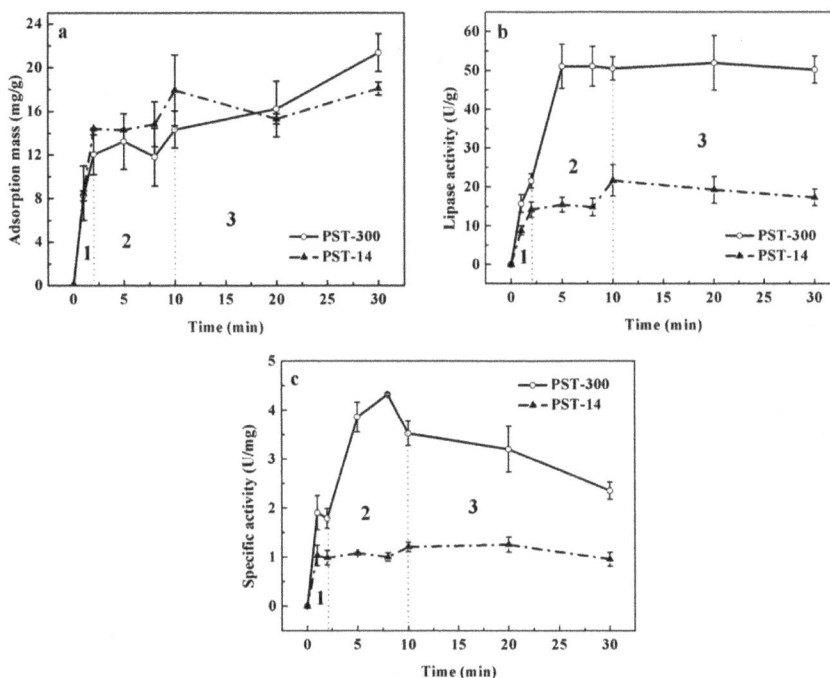

Figure 3. Adsorption of lipase (0.05 mg/mL) into PST microspheres within short time ((a) Adsorption mass; (b) Lipase activity; (c) Specific activity). Specific activity of free CALB: 3.8 U/mg.

In Figure 3a, the adsorption mass increased rapidly in the first two minutes, and monolayer adsorption mainly occurred in this stage since the enzyme concentration was very low (0.05 mg/mL). For the gigaporous microspheres, the adsorption mass basically maintained a sustained growth from the time point of 2 to 30 min. As for mesoporous microspheres, the adsorption mass increased from 12.2 to 15.6 mg/g (from the time point of 2 to 10 min), and nearly kept stable after the time point of 10 min. Therefore, after the time point of 10 min, the adsorption was mainly multilayer. Figures 4 and 5 were the CLSM images and the change of fluorescence intensity of lipase adsorbed in gigaporous and mesoporous microspheres with the adsorption time. For the gigaporous particles, before the time point of 2 min, lipase molecules entered the whole particle and dispersed homogeneously (Figure 4a), while, in the mesoporous particles, most enzymes adsorbed in the shell (Figure 5a). Consistent with the increase of adsorption mass in Figure 3a, after the time point of 2 min, the fluorescence intensity in gigaporous particles exhibited a stable and uniform enhance (Figure 4b–d). As for lipase adsorbed in the mesoporous microspheres, only the fluorescence intensity of shell increased due to the limit of a smaller pore size (Figure 5b–d).

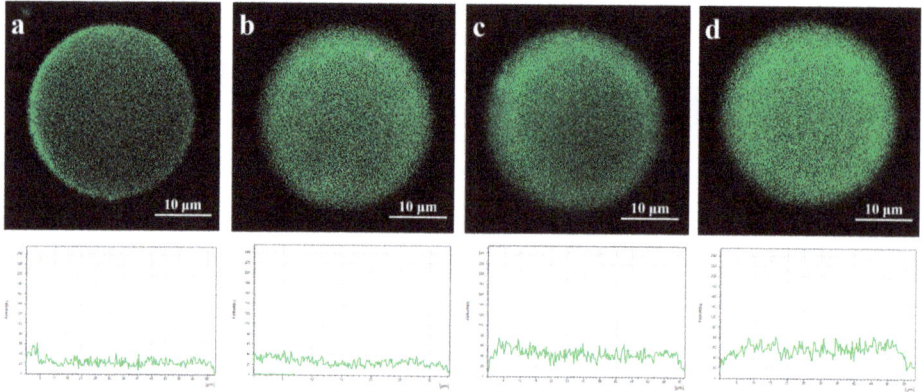

Figure 4. CLSM images and fluorescence distribution of lipase adsorbed in gigaporous microspheres at different time point ((**a**) 2 min; (**b**) 8 min; (**c**) 30 min; (**d**) 40 min).

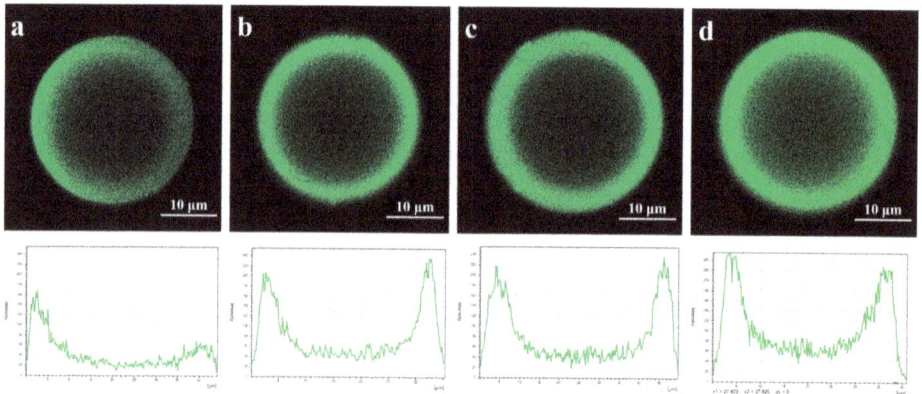

Figure 5. CLSM images and fluorescence distribution of lipase adsorbed in mesoporous microspheres at different time point ((**a**) 2 min; (**b**) 8 min; (**c**) 30 min; (**d**) 40 min).

The effects of the different adsorption behavior in the carriers with various porous structures were researched. The dramatic difference between their enzyme activities (Figure 3b), especially the specific activities (Figure 3c) gave very interesting results. In the first stage (before the time point of 2 min), the specific activity of lipase immobilized in the gigaporous particles was 1.79 U/mg, and that in the mesoporous particles was only 0.99 U/mg. According to the fluorescence distribution, the specific area of the gigaporous and mesoporous particles, and the area per lipase molecule occupied in ideal state, we calculated the amount of enzyme adsorbed on the unit area of the microspheres, 5.34×10^{11} and 3.12×10^{10}, respectively. The amount is far less than the molecule number needed for monolayer adsorption in both type of particles (7.85×10^{14}, PST-300; 7.68×10^{15}, PST-14, 1/10 diameter). Therefore, it tended to be monolayer adsorption in this stage theoretically. Since both kinds of microsphere were based on P(ST-DVB) [poly(styrene-divinylbenzene)], which allowed physical adsorption of lipase, the only difference is the curvature of pores in the stage of monolayer adsorption. The curvature of gigapores was much lower, and the surface of gigapores was flatter than that of mesopores. According to the effect of curvature to proteins [27], when the value of curvature was bigger, the contact area and the deformation were larger. Additionally, the structure of the enzyme molecules cannot be well maintained. Conversely, the flatter surface was beneficial for retaining the molecular structure. This explained the difference of specific activity in gigaporous and mesoporous particles before the time point of 2 min.

For lipase immobilized in gigaporous particles, from the time point of 2 to 8 min in Figure 3c, their specific activity increased rapidly to 3.86 U/mg. While the specific activity of lipase immobilized in mesoporous particles did not have significant change. The main reason of the difference is also due to the pore structure. From the time point of 2 min, multilayer adsorption occurred in gigapores with the large enough space. Multilayer adsorption effectively maintained the enzyme activity. The adsorption of lipase in mesoporous microspheres was still inclined to be monolayer adsorption because the mesopores restricted the distribution of lipase molecules (the pore size is 14.7 nm; the molecular size of CALB is $6.92 \times 5.05 \times 8.67$ nm^3 [28]). Since the monolayer adsorption of lipase was dominant in mesoporous microspheres, the enzyme activity and the specific activity were obviously lower than those in the gigaporous ones.

After the time point of 8 min, though the adsorption mass still increased in the gigaporous microspheres, the enzyme activity had little change, and the specific activity decreased. This was presumably due to the fact that with more enzyme adsorbed in the pores, the cross-section of the pores might be reduced; thus, the diffusion resistance for the substrate and the product was increased [29].

3.2. Comparison of the Adsorption in Microsphere with the Adsorption on Flat Surface

The reasons for high enzyme activity and specific activity of lipase in gigaporous microspheres are explained by the difference of the curvature and the monolayer/multilayer adsorption. The adsorption on a flat surface was further compared to find what happened when there was no limit of pores. Quartz crystal microbalance with dissipation monitoring (QCM-D) technology was adopted to investigate the adsorption behavior of lipase on PST chips and the influence on enzyme activity. Figure 6 showed the adsorption parameters of lipase on PST chips.

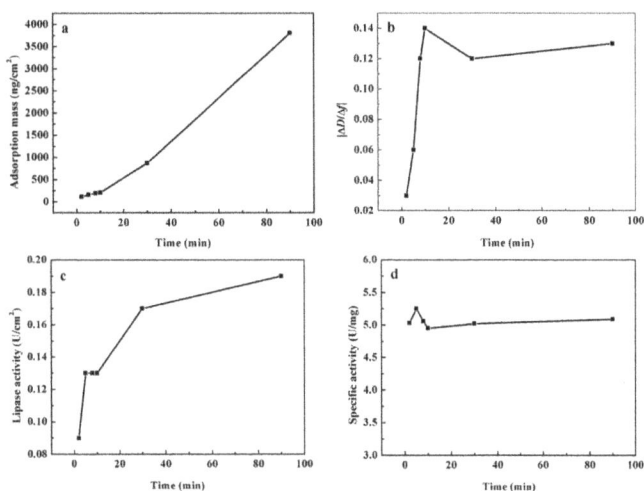

Figure 6. Adsorption of lipase on PST chip determined by QCM-D ((**a**) Adsorption mass; (**b**) $|\Delta D/\Delta f|$ ratio; (**c**) Lipase activity; (**d**) Specific activity).

From Figure 6a, we could see that the increase of adsorption mass was a rough linear rise on the chip when excluding the influence of diffusion transfer in pores. The change of the $|\Delta D/\Delta f|$ ratio in Figure 6b reflected the tightness degree of lipase adsorbed on the chip. At the beginning, lipase molecules mainly interacted with the surface of the chip, and the interaction was strong. Therefore, the $|\Delta D/\Delta f|$ ratio was very low. With the increase of the adsorption mass, the enzyme layers got thicker, and the $|\Delta D/\Delta f|$ ratio increased rapidly. After the time point of 10 min, more multilayer adsorption of enzyme molecules enhanced the tightness of the layers, and the $|\Delta D/\Delta f|$ ratio then became stable.

From Figure 6c,d, we could see that the enzyme activity increased with the increasing adsorption amount, but the specific activity had little change, which is dramatically different from that of gigaporous particles. For the adsorption on the chip, the effect of pores was completely ruled out. However, for gigaporous particles, though most pores distributed above 100 nm, there were still so many small pores in which the diffusion resistance and the structure change of lipase existed, which could reduce the specific activity of lipase.

4. Conclusions

The differences of mesoporous particles, gigaporous particles, and a flat surface for enzyme immobilization were investigated in this study. It was found that the adsorption of enzyme in the mesoporous particles was inclined to be monolayer adsorption that may change the enzyme structure, and the small pores limited the mass transfer of the substrate and the product. These factors reduced the specific activity of lipase. As for the gigaporous particles, in the stage of monolayer adsorption, the interaction between lipase molecules and the pores was weaker than that in the mesopores, so the specific activity was higher. With the increasing amount of enzyme, the occurrence of multilayer adsorption could further promote the maintenance of the enzyme activity. When lipase adsorbed on a chip, the behavior was quite different when excluding the influence of pores, the adsorption mass showed a linear increase, the specific activity had little change during the whole process, and the activity continuously enhanced.

Acknowledgments: The authors acknowledge the financial support by National Natural Science Foundation of China (No. 21336010, No. 21106161), the National Science and Technology Support Program (No. 2012BAD32B09), and 863 Project (2012AA02A406).

Author Contributions: Weichen Wang and Weiqing Zhou conceived and designed the experiments and analyzed the data; Wei Wei and Dongxia Hao helped to modify the manuscript; Juan Li helped with the preparation of the microspheres; Zhiguo Su provided many constructive suggestions about this work; Guanghui Ma provided suggestions and supervised the work and the manuscript editing.

Conflicts of Interest: The authors declare no conflict of interest.

Abbreviations

The following abbreviations are used in this manuscript:

PST	polystyrene
CALB	lipase B from *Candida antarctica*
CLSM	confocal laser scanning microscopy
QCM-D	quartz crystal microbalance with dissipation monitoring

References

1. Pinto, M.C.C.; Freire, D.M.G.; Pinto, J.C. Influence of the morphology of core-shell supports on the immobilization of lipase B from *Candida antarctica*. *Molecules* **2014**, *19*, 12509–12530. [CrossRef] [PubMed]
2. Zhang, C.; Luo, S.; Chen, W. Activity of catalase adsorbed to carbon nanotubes: Effects of carbon nanotube surface properties. *Talanta* **2013**, *113*, 142–147. [CrossRef] [PubMed]
3. Gustafsson, H.; Johansson, E.M.; Barrabino, A.; Odén, M.; Holmberg, K. Immobilization of lipase from *Mucor miehei* and *Rhizopus oryzae* into mesoporous silica—The effect of varied particle size and morphology. *Colloids Surf. B Biointerfaces* **2012**, *100*, 22–30. [CrossRef] [PubMed]

4. Sabbani, S.; Hedenström, E.; Nordin, O. The enantioselectivity of *Candida rugosa* lipase is influenced by the particle size of the immobilising support material Accurel. *J. Mol. Catal. B Enzym.* **2006**, *42*, 1–9. [CrossRef]
5. Gao, S.; Wang, Y.; Diao, X.; Luo, G.; Dai, Y. Effect of pore diameter and cross-linking method on the immobilization efficiency of *Candida rugosa* lipase in SBA-15. *Bioresource Technol.* **2010**, *101*, 3830–3837. [CrossRef] [PubMed]
6. Chaijitrsakool, T.; Tonanon, N.; Tanthapanichakoon, W.; Tamon, H.; Prichanont, S. Effects of pore characters of mesoporous resorcinol–formaldehyde carbon gels on enzyme immobilization. *J. Mol. Catal. B Enzym.* **2008**, *55*, 137–141. [CrossRef]
7. Liu, T.; Liu, Y.; Wang, X.; Li, Q.; Wang, J.; Yan, Y. Improving catalytic performance of *Burkholderia cepacia* lipase immobilized on macroporous resin NKA. *J. Mol. Catal. B Enzym.* **2011**, *71*, 45–50. [CrossRef]
8. Kang, Y.; He, J.; Guo, X.; Guo, X.; Song, Z. Influence of pore diameters on the immobilization of lipase in SBA-15. *Ind. Eng. Chem. Res.* **2007**, *46*, 4474–4479. [CrossRef]
9. Zhou, Z.; Hartmann, M. Progress in enzyme immobilization in ordered mesoporous materials and related applications. *Chem. Soc. Rev.* **2013**, *42*, 3894–3912. [CrossRef] [PubMed]
10. Hartmann, M.; Kostrov, X. Immobilization of enzymes on porous silicas—benefits and challenges. *Chem. Soc. Rev.* **2013**, *42*, 6277–6289. [CrossRef] [PubMed]
11. Bayne, L.; Ulijn, R.V.; Halling, P.J. Effect of pore size on the performance of immobilised enzymes. *Chem. Soc. Rev.* **2013**, *42*, 9000–9010. [CrossRef] [PubMed]
12. Fried, D.I.; Brieler, F.J.; Froeba, M. Designing inorganic porous materials for enzyme adsorption and applications in biocatalysis. *ChemCatChem* **2013**, *5*, 862–884. [CrossRef]
13. Schlipf, D.M.; Rankin, S.E.; Knutson, B.L. Pore-size dependent protein adsorption and protection from proteolytic hydrolysis in tailored mesoporous silica particles. *Acs. Appl. Mater. Interfaces* **2013**, *5*, 10111–10117. [CrossRef] [PubMed]
14. Vijayaraj, M.; Gadiou, R.; Anselme, K.; Ghimbeu, C.; Vix-Guterl, C.; Orikasa, H.; Kyotani, T.; Ittisanronnachai, S. The influence of surface chemistry and pore size on the adsorption of proteins on nanostructured carbon materials. *Adv. Funct. Mater.* **2010**, *20*, 2489–2499. [CrossRef]
15. Takahashi, H.; Li, B.; Sasaki, T.; Miyazaki, C.; Kajino, T.; Inagaki, S. Catalytic activity in organic solvents and stability of immobilized enzymes depend on the pore size and surface characteristics of mesoporous silica. *Chem. Mater.* **2000**, *12*, 3301–3305. [CrossRef]
16. Weber, E.; Sirim, D.; Schreiber, T.; Thomas, B.; Pleiss, J.; Hunger, M.; Gläser, R.; Urlacher, V.B. Immobilization of P450 BM-3 monooxygenase on mesoporous molecular sieves with different pore diameters. *J. Mol. Catal. B Enzym.* **2010**, *64*, 29–37. [CrossRef]
17. Li, Y.; Gao, F.; Wei, W.; Qu, J.B.; Ma, G.H.; Zhou, W.Q. Pore size of macroporous polystyrene microspheres affects lipase immobilization. *J. Mol. Catal. B Enzym.* **2010**, *66*, 182–189. [CrossRef]
18. Mei, Y.; Miller, L.; Gao, W.; Gross, R.A. Imaging the distribution and secondary structure of immobilized enzymes using infrared microspectroscopy. *Biomacromolecules* **2003**, *4*, 70–74. [CrossRef] [PubMed]
19. Loos, K.; Kennedy, S.B.; Eidelman, N.; Tai, Y.; Zharnikov, M.; Amis, E.J.; Ulman, A.; Gross, R.A. Combinatorial approach to study enzyme/surface interactions. *Langmuir* **2005**, *21*, 5237–5241. [CrossRef] [PubMed]
20. Chen, B.; Miller, E.M.; Miller, L.; Maikner, J.J.; Gross, R.A. Effects of macroporous resin size on *Candida antarctica* lipase B adsorption, fraction of active molecules, and catalytic activity for polyester synthesis. *Langmuir* **2007**, *23*, 1381–1387. [CrossRef] [PubMed]
21. Chen, B.; Miller, E.M.; Gross, R.A. Effects of porous polystyrene resin parameters on *Candida antarctica* lipase B adsorption, distribution, and polyester synthesis activity. *Langmuir* **2007**, *23*, 6467–6474. [CrossRef] [PubMed]
22. Miletić, N.; Vuković, Z.; Nastasović, A.; Loos, K. Macroporous poly(glycidyl methacrylate-*co*-ethylene glycol dimethacrylate) resins—Versatile immobilization supports for biocatalysts. *J. Mol. Catal. B Enzym.* **2009**, *56*, 196–201. [CrossRef]
23. Nihira, T.; Mori, T.; Asakura, M.; Okahata, Y. Kinetic studies of dextransucrase enzyme reactions on a substrate- or enzyme-immobilized 27 MHz quartz crystal microbalance. *Langmuir* **2011**, *27*, 2107–2111. [CrossRef] [PubMed]
24. Smith, P.; Krohn, R.I.; Hermanson, G.; Mallia, A.; Gartner, F.; Provenzano, M.; Fujimoto, E.; Goeke, N.; Olson, B.; Klenk, D. Measurement of protein using bicinchoninic acid. *Anal. Biochem.* **1985**, *150*, 76–85. [CrossRef]

25. Su, X.; Zong, Y.; Richter, R.; Knoll, W. Enzyme immobilization on poly(ethylene-*co*-acrylic acid) films studied by quartz crystal microbalance with dissipation monitoring. *J. Colloid Interface Sci.* **2005**, *287*, 35–42. [CrossRef] [PubMed]

26. Rodahl, M.; Kasemo, B. On the measurement of thin liquid overlayers with the quartz-crystal microbalance. *Sens. Actuators A Phys.* **1996**, *54*, 448–456. [CrossRef]

27. Hao, D.-X.; Sandström, C.; Huang, Y.-D.; Kenne, L.; Janson, J.-C.; Ma, G.-H.; Su, Z.-G. Residue-level elucidation of the ligand-induced protein binding on phenyl-argarose microspheres by NMR hydrogen/deuterium exchange technique. *Soft Matter* **2012**, *8*, 6248–6255. [CrossRef]

28. Cao, L. *Carrier-Bound Immobilized Enzymes: Principles, Application and Design*, 1st ed.; Wiley-VCH: New York, NY, USA, 2006.

29. Geitmann, M.; Danielson, U.H. Studies of substrate-induced conformational changes in human cytomegalovirus protease using optical biosensor technology. *Anal. Biochem.* **2004**, *332*, 203–214. [CrossRef] [PubMed]

polymers

MDPI

Article

Green Synthesis of Smart Metal/Polymer Nanocomposite Particles and Their Tuneable Catalytic Activities

Noel Peter Bengzon Tan, Cheng Hao Lee and Pei Li *

Department of Applied Biology and Chemical Technology, The Hong Kong Polytechnic University, Hung Hom, Kowloon, Hong Kong, China; bengzontan@nami.org.hk (N.P.B.T.); chenghao.lee@polyu.edu.hk (C.H.L.)
* Correspondence: pei.li@polyu.edu.hk; Tel.: +852-3400-8721

Academic Editor: Haruma Kawaguchi
Received: 17 February 2016; Accepted: 17 March 2016; Published: 23 March 2016

Abstract: Herein we report a simple and green synthesis of smart Au and Ag@Au nanocomposite particles using poly(N-isopropylacrylamide)/polyethyleneimine (PNIPAm/PEI) core-shell microgels as dual reductant and templates in an aqueous system. The nanocomposite particles were synthesized through a spontaneous reduction of tetrachloroauric (III) acid to gold nanoparticles at room temperature, and *in situ* encapsulation and stabilization of the resultant gold nanoparticles (AuNPs) with amine-rich PEI shells. The preformed gold nanoparticles then acted as seed nanoparticles for further generation of Ag@Au bimetallic nanoparticles within the microgel templates at 60 °C. These nanocomposite particles were characterized by TEM, AFM, XPS, UV-vis spectroscopy, *zeta*-potential, and particle size analysis. The synergistic effects of the smart nanocomposite particles were studied via the reduction of *p*-nitrophenol to *p*-aminophenol. The catalytic performance of the bimetallic Ag@Au nanocomposite particles was 25-fold higher than that of the monometallic Au nanoparticles. Finally, the controllable catalytic activities of the Au@PNIPAm/PEI nanocomposite particles were demonstrated via tuning the solution pH and temperature.

Keywords: gold nanoparticles; silver/gold bimetallic nanoparticles; smart core-shell microgel; metal/polymer nanocomposite; reduction of *p*-nitrophenol

1. Introduction

Metal nanoparticles have gained much attention over the past two decades because of their unique properties. Their distinctive size and shape-dependent optical and electronic properties have opened up vast potential applications in various fields [1]. However, metallic nanoparticles have a tendency to aggregate due to their large surface energy, which drives the thermodynamically favored coalescence process. Various types of stabilizing agents have been used to prevent nanoparticles from aggregating. They include ligands [2], surfactants [3,4], polymers [5,6], dendrimers [7], cyclodextrin [8], and microgels [9]. Recently, the use of smart polymer particles to stabilize metallic nanoparticles has received increasing attention because of the synergistic properties of the smart metal/polymer nanocomposite particles, derived from responsive polymer and metallic nanoparticles [10,11]. Smart metal/polymer nanocomposite particles of this kind have shown promising potential for use as hybrid materials for biomedical applications [12,13], sensing [14], and catalytic reactions [15,16].

Our group has recently reported a simple and green synthesis of AuNPs/polymer nanocomposite particles through a spontaneous reduction of tetrachloroauric (III) acid and encapsulation of the resultant gold nanoparticles using amine-rich core-shell particles in water [17,18]. The high local amine concentration of the polyethyleneimine shell enables effective reduction of gold ions to gold nanoparticles in the absence of any organic solvents, reducing agents, or stabilizers. Thus we envision that smart metal/polymer nanocomposite particles synthesized by this green approach may find potential application in catalytic reactions because the microgel particles can not only stabilize the metal nanoparticles and retain their nanoscale properties, but also allow the switching on and off of the catalytic activity of the metal nanoparticles through controlling the accessibility of the metal nanoparticles to the reactant. Moreover, the nanocomposite particles can be easily recovered and reused. The nanocomposite particles containing bimetallic nanoparticles are of particular interest because of their possible synergistic effects [19]. For example, Ahn and co-workers have recently reported the synthesis of Ag@Au bimetallic nanoparticles on magnetic silica microspheres through seeding, coalescing, seeds to cores, and then growing shells from the cores on aminopropyl functionalized silica microspheres [20]. Xin *et al.* have also synthesized Au@Ag bimetallic nanoparticles using polyelectrolyte multilayers (PEMs) of poly(styrene sulfonate) (PSS) and poly(diallyldimethylammonium chloride) (PDDA) particles as support [21]. Electrostatic interaction between the cations and anions of the PEMs and metal ions was used as a driving force to attract metal ions into the system. Subsequent reduction of metal salts with reducing agents such as $NaBH_4$ and ascorbic acid was carried out in repeated cycles to improve the loading and size of the metal nanoparticles. However, current methods for generating bimetallic nanoparticles usually involve tedious procedures that are not amenable for scale-up production. In this study, we have synthesized a dual responsive microgel particle that consists of a temperature-sensitive core of poly(N-isopropylacryamide) and a pH-sensitive polyethyleneimine shell.. This type of pH- and temperature-responsive microgel particles was used to generate silver in gold Ag@Au/polymer nanocomposite particles. The synergistic effects of the smart nanocomposite particles were studied using the reduction of *p*-nitrophenol to *p*-aminophenol as a model reaction. The controllable catalytic activity of the Au@PNIPAm/PEI nanocomposite particles was demonstrated via the tuning of solution pH and temperature.

2. Materials and Methods

2.1. Materials

Spindle-crystals of N-isopropylacrylamide (NIPAM, Sigma-Aldrich, Saint Louis, MO, USA) were purified by repeated recrystallization in a mixture of toluene and n-hexane (1:5 *v/v*). Branched polyethyleneimine (PEI) with an average molecular weight of 750,000 (50 wt % solution in water), N,N-methylenebisacrylamide (MBA) and *tert*-butyl hydroperoxide (TBHP, 70% aqueous solution), hydrogen tetrachloroaurate (III) trihydrate (HAuCl4· 3H2O), silver nitrate (AgNO3), *p*-nitrophenol (reagent grade), and sodium borohydride were all purchased from Sigma-Aldrich Chemical Co., and used as received. Deionized water or Milli-Q water was used for dilution and dispersion medium.

2.2. Synthesis of Au@PNIPAm/PEI Nanocomposite Particles

The synthesis of Au-loaded poly(N-isopropyl acrylamide)/polyethyleneimine (PNIPAm/PEI) nanocomposite particles was based on our previously established method [17]. Such a method involved a simple mixing of preformed PNIPAm/PEI microgel particles with a hydrogen tetrachloroaurate (III) trihydrate (HAuCl4· 3H2O) solution according to the following general procedure: A stock solution of hydrogen tetrachloroaurate (III) trihydrate (1.317×10^{-3} M) was purged with N2 for 30 min. It was then added dropwise (1 mL) to the PNIPAm/PEI microgel dispersion (20 mL, 400 ppm, molar ratio of N/Au^{3+} = 28). The mixture was stirred at 250 rpm for 4 h at 25 °C. The resultant gold-loaded microgels were purified by a single cycle of centrifugation at 12,000 rpm and 10 °C for 1 h. The collected pink product was redispersed in deionized water under sonication for subsequent usage.

2.3. Synthesis of Ag@Au/PNIPAm/PEI Nanocomposite Particles

Synthesis of the Ag@Au bimetallic nanoparticles was carried out through a successive reduction of AgNO3 in the presence of the Au@PNIPAm/PEI particles. The preformed gold nanoparticles were used as seeds for the successive reduction of the silver ions to silver nanoparticles. A 1:1 molar ratio sample that contained 1 mL Au salt solution (1.317×10^{-3} M) and 1 mL Ag salt solution (1.317×10^{-3} M) was used in this study. The reaction was first carried out at room temperature for 30 min, followed by heating at 60 °C for another 30 min. During the mixing of the silver salt solution with the Au-loaded particles, the solution changed color from light pink to light gray, and eventually turned fully gray after heating the mixture.

2.4. Catalytic Activity of Nanocomposite Particles

The catalytic activity of both Au@PNIPAm/PEI and Ag@Au/PNIPAm/PEI nanocomposite particles were studied in an aqueous system using reduction of *p*-nitrophenol to *p*-aminophenol as a model reaction. Sodium borohydride (200 μL, 0.001 M), *p*-nitrophenol (30 μL, 0.001 M) and a definite quantity of the composite particles (40 μL, 1.5 wt % solid content) were diluted to 2 mL with deionized water and mixed in a cuvette reactor. The reaction mixture changed from light yellow to colorless. The catalytic reaction using Au@PNIPAm/PEI (amine to gold molar ratio of 28:1) and Ag@Au/PNIPAm/PEI [amine to gold and silver ions ratio ($N/Au^{3+}Ag^+$) of 28:1] were systematically studied under different solution pH values (3–11) and temperatures (25–39 °C). The reduction of *p*-nitrophenol to *p*-aminophenol was monitored by the UV-visible spectroscopy (Agilent Technologies, Santa Clara, CA, USA) at a wavelength of 400 nm for 20 min. Measurements were conducted at 2-min intervals.

2.5. Measurements and Characterization

2.5.1. Particle Sizes and Surface Charges

The hydrodynamic diameter and *zeta*-potential of the microgel templates as well as mono- and bimetallic-loaded nanocomposite particles were all measured with a Beckman Coulter Delsa Nano particle analyzer (Beckman Coulter, Brea, CA, USA) using a photon correlation spectroscopy with electrophoretic dynamic light scattering (a two-laser diode light source with a wavelength of 658 nm at 30 mW). Hydrodynamic diameter, D_h, was obtained from the Einstein Stokes equation, $D_h = kT/3\pi\eta D$, where k is the Boltzmann constant, η is the dispersant viscosity, T is the temperature (K), and D is the diffusion coefficient. The diffusion coefficient was obtained from the decay rate of the intensity correlation function of the scattered light (*i.e.*, correlogram), $G(\tau) = \int I(t)I(t+\tau)dt$. Each measurement was carried out in triplicate. *zeta*-Potential measures the surface charge of particles based on their electrophoretic mobility. Samples for *zeta*–potential measurements were diluted to 100–200 ppm with 1 mM NaCl and measured at 25 °C.

2.5.2. Transmission Electron Microscopy

Transmission electron microscopy (TEM) images of monometallic gold-loaded microgels and bimetallic (Ag@Au)-loaded microgels were observed using a transmission electron microscope (JEOL 100 CX, JEOL, Tokyo, Japan) at an accelerating voltage of 100 kV. The high resolution TEM micrographs of Au and Ag@Au nanoparticles and their corresponding selected area electron diffraction (SAED) patterns were characterized by a JEOL 2010 TEM (JEOL) at an accelerating voltage of 200 kV. The sample was prepared by wetting a carbon-coated grid with a 5 μL of the diluted particle dispersion, followed by drying at room temperature prior to TEM analysis. There was no pretreatment staining for all nanocomposite samples.

2.5.3. X-ray Photoelectron Spectroscopy

X-ray photoelectron spectroscopy (XPS) data were recorded on a multi-surface analysis system (PHI 5600, Physical Electronics, Chanhassen, MN, USA) with a monochromatic AlK$_\alpha$ X-ray source (1486.6 eV). Sample spot sizes varied from 200 to 400 μm in diameter. The pass energies of exciting radiations were set at 187 and 45 eV for survey and elemental scans, respectively. The energy and emission currents of the electrons were 4 eV and 0.35 mA, respectively. Energy resolution was at 0.7 eV with a chamber pressure of 5×10^{-10} torr. Spectral calibration was determined by setting the C 1s component at 285.0 eV. All data acquisition was processed with a PC-based Advantage software (version 1.85, Advantage Software Co., Stuart, FL, USA). The surface composition was determined by using the manufacturer's sensitivity factors. Curve fitting of the spectrum was accomplished using a nonlinear least-squares method. A Gaussian function was assumed for the curve fitting. The deconvolution of carbon, oxygen, and nitrogen peaks was processed with MagicPlot software (version 2.5.1, MagicPlot Systems, St. Petersburg, Russia).

2.5.4. UV-vis Spectroscopy

UV-vis spectra were recorded on a Varian Cary 4000 Spectrophotometer using wavelengths ranging from 250–525 nm with an absorbance set from 0 to 1.40 a.u. Samples were diluted to appropriate concentrations and measured in the 5-mL cuvette. Actual absorbance as a function of time was plotted and a fitted curve was derived from an analysis-fitting function of Origin Pro software (v. 8.0, OriginLab Co., Northampton, MA, USA). Time-dependent UV-vis spectra (Agilent Technologies) on the reduction of *p*-nitrophenol to *p*-aminophenol using both monometallic (Au@PNIPAm/PEI) and bimetallic (Ag@Au/(PNIPAm/PEI)) nanocomposite particles as catalysts were used to monitor the catalytic reaction and calculate the catalytic constant, *k*. This constant is proportional to the catalytic rate of the reaction. UV-vis absorbance at 400 nm is a characteristic peak of the *p*-nitrophenol. During the course of the catalytic reaction, the intensity of this peak decreased due to its conversion to *p*-aminophenol, which appeared at wavelengths between 290 and 310 nm. The gradual reduction in UV-vis absorbance of the *p*-nitrophenol was monitored at specific time intervals until its full disappearance.

2.5.5. Atomic Force Microscopy

Atomic force microscopy (AFM) images of PNIPAm/PEI microgel particles at 29 and 45 °C in aqueous solution were obtained using XE-120 inverted microscope complete AFM system with universal liquid cell option (Park Systems, Suwon, Korea). Imaging was carried out in a temperature control stage using a standard silicon nitride (Si_3N_4), gold-coated cantilever tip (MLCT-AUHW, Veeco, Plainview, NY, USA) in a non-contact fluid. The AFM system carries a decoupled XY-scanner with a maximum scan range of 100 μm × 100 μm. Image matrix was 256 × 256 pixels for fluid samples. Scan rate used was 0.5 to 0.8 Hz depending on the sample conditions. Humidity was adjusted from 40% to 80%.

2.5.6. Elemental Analysis of Bimetallic Nanoparticles

The atomic percentage of individual bimetallic nanoparticle was evaluated using an Energy-dispersive Spectrometry (EDS) probe, operating in the bright field mode of a JEOL 2010 TEM. The spectral resolution was 1 nm. In the chemical composition measurement, Au showed energy intensities at 2 and 2.6 keV, whereas Ag intensities were at 3 and 3.4 keV.

3. Results and Discussion

3.1. Synthesis of Au and Ag@Au/PNIPAm/PEI Nanocomposite Particles

The synthesis of both mono- and bimetallic nanoparticles is illustrated in Scheme 1. The gold nanoparticles were first generated through a reduction of gold salt with a polymeric amine using core-shell particles that consisted of poly(N-isopropyl acrylamide) cores and polyethyleneimine shells. Generation and stabilization mechanisms of gold nanoparticles and formation of Au@PNIPAm/PEI nanocomposite particles have been discussed in our previous papers [17,18]. Amine groups are known to have a reducing ability to generate metal nanoparticles [22]. They also can complex with metal ions and metal nanoparticles through their chelating properties [23,24]. When using the PNIPAm/PEI core-shell template, gold salt ions $[AuCl_4]^-$ were attracted into the template through an electrostatic interaction between the positively charged PEI shell and the negatively charged gold salt ions. The high local amine concentration could significantly enhance the reduction rate of gold salt ions to generate gold nanoparticles without the aid of any reducing agents.

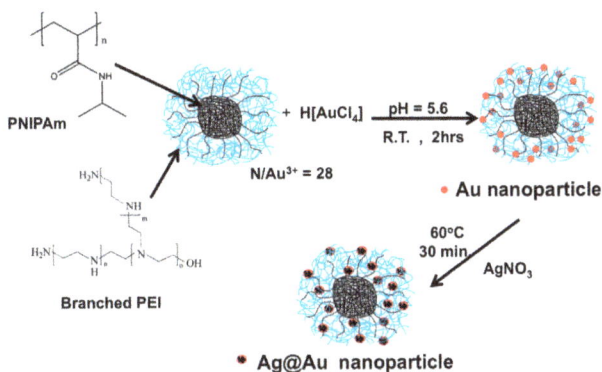

Scheme 1. Reaction scheme of the formation of Au and Ag@Au metal nanoparticles using PNIPAm/PEI template.

Formation of the AuNPs was evident from the change in solution color from white to light pink, which occurred after 30 to 40 min of reaction at room temperature. These gold nanoparticles were then used as seeds for the successive reduction of silver ions to bimetallic Ag/Au nanoparticles using silver nitrate solution. The reduction was carried out at 60 °C in order to increase the conversion and crystallinity of the bimetallic nanoparticles. In the presence of gold metal nanoparticles, silver ions (Ag^+) could be reduced to silver nanoparticles via an under-potential deposition mechanism [25], or, as others refer to it, the noble metal induced reduction (NMIR) method [26]. In this mechanism, the gold nanoparticles acted as seeds or active sites for further growth of silver nanoparticles. The formation of silver in gold bimetallic nanoparticles was evident upon a change in the solution color from light pink to gray. This effect is attributed to the fact that the ionization potential and electron affinity values of Au atom are higher than those of the Ag atom. The large electronegativity value of the Au atom leads to effective charge transfer from silver to gold atoms because the second metal ion (Ag^+) has lower reduction potential than gold nanoparticles [27,28].

3.2. Compositions and Morphologies of Au and Ag@Au/PNIPAm/PEI Nanocomposite Particles

The TEM images shown in Figure 1 reveal the transformation of PNIPAm/PEI to Au and Ag@Au/PNIPAm/PEI nanocomposite particles. The PNIPAm/PEI particle displays a core-shell structure where the core has a darker contrast than the shell (Figure 1a inset). The resultant Au@PNIPAm/PEI particles are shown in Figure 1b. The gold nanoparticles look like clusters of small gold nanoparticles with an average diameter of 15 ± 4.0 nm. They were homogenously distributed within the particles. Figure 1c shows the morphology of Ag@AuPNIPAm/PEI nanocomposite particles. The bimetallic nanoparticles displayed two different intensities of contrast (Supplementary Materials Figure S1). The region in a darker shade of gray constitutes the gold nanoparticle while the lighter gray part is the silver nanoparticles. Elemental analysis results revealed that the bimetallic nanoparticles comprised 17.6 and 82.4 atomic percentage of Ag and Au, respectively (Supplementary Materials Figure S2). The higher Au percentage may be attributed to the fact that only partial Au^{3+} ions were initially reduced to gold nanoparticles at room temperature, while the remaining gold salt ions $[AuCl_4]^-$ were attracted to the positively charged PEI shells. When the temperature was raised to 60 °C, the other part of the gold salt ions was further reduced together with some Ag^+ ions, resulting in the formation of Ag@Au nanoparticles with a higher Au content.

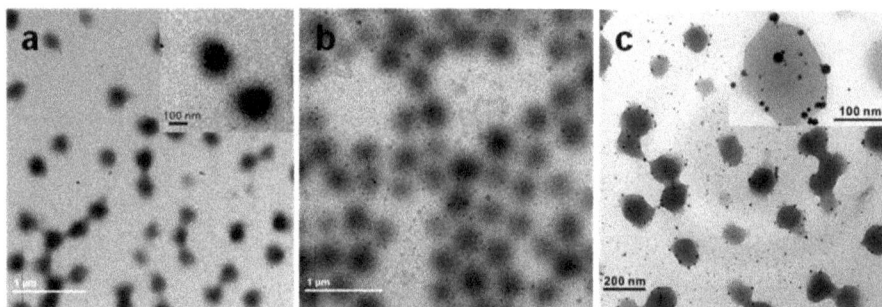

Figure 1. TEM images of (**a**) PNIPAm/PEI microgel templates; (**b**) Au/PNIPAm/PEI nanocomposite particles prepared at room temperature; (**c**) Ag@Au/PNIPAm/PEI nanocomposite particles formed at 60 °C (Au^{3+}/Ag$^+$ molar ratio of 1:1).

It was noted that the average size of the bimetallic nanoparticles was only *ca.* 6.83 ± 2.5 nm as shown in Figure 1c. Their sizes were much smaller than those of the original Au nanoclusters, indicating that the Au nanoclusters were dissociated after forming bimetallic nanoparticles. The smaller bimetallic nanoparticles were more difficult to entrap within the soft and flexible PEI shell. Thus some of them escaped from the templates and dispersed in the solution. Furthermore, the bimetallic nanoparticles located in the templates or freely dispersed in solution appeared much darker than the Au nanoclusters. This phenomenon may be attributed to the thermal treatment of the nanocomposite particles, a process that facilitates the growth of the metallic nanoparticles through further crystallization.

A high-resolution TEM (HRTEM) was performed to verify the nanostructures of mono- and bimetallic nanoparticles. In order to clearly observe the metal nanoparticles in the image, the nanocomposite particles were pre-treated with electron beam irradiation with a current density of 0.4 nA/nm^2 for 30 s to partially remove the polymer template. Figure 2a shows an image of a polycrystalline Au nanostructure with a diameter of *ca.* 19.1 nm. Figure 2b shows the selected area electron diffraction (SAED) pattern over several Au nanoparticles. It reveals a ring pattern indexed as (111), (200), (220), (311), and (331) of a face-centered cubic (fcc) gold lattice. Thus, the gold nanoparticle is mainly composed of (111) planes with a *d*-spacing of 0.236 nm. The fuzzy central portion of the nanoparticle exhibited in the HRTEM image indicates that the Au nanoparticle adopts an icosahedral morphology with multiple-twinned structure [29]. Such defect is attributed to the small displacement of atoms located at the central nucleation site of the nanoparticle. Figure 2c,d show the crystallinity and different lattice arrangement of the Ag@Au bimetallic nanoparticles, respectively. The difference in electron densities between gold and silver is due to their differences in the corresponding lattice parameter. Au (111) has a lattice *d*-spacing of 0.236 nm, while that of Ag (200) is 0.205 nm. Hence, the SAED analysis of the bimetallic nanoparticles further confirmed the co-existence of the crystalline orientations of Au and Ag in a single nanoparticle.

3.3. Particle Sizes and Surface Charges of the Nanocomposite Particles

Changes in particle sizes and surface charges of the original PNIPAm/PEI template, the Au@PNIPAm/PEI, and the Ag@Au/PNIPAm/PEI nanocomposite particles have been examined by measuring their hydrodynamic sizes and *zeta*-potential values. The results in Table 1 show that the introduction of gold ions into the PNIPAm/PEI template decreases both the particle size and the surface charge. The addition of Ag$^+$ ions to form bimetallic nanocomposite particles further decreases both the particle size and the surface charge. These effects may be attributed to the strong binding affinity of the Au or Ag@Au nanoparticles with the amine groups, resulting in the contraction of a highly swollen PEI shell. The conversion of hydrophilic amino groups to more hydrophobic moieties or crosslinking of the PEI chains due to the formation of amine radical cation during the gold ion reduction is another possible reason [22]. Despite the decrease of surface charges to almost +5 mV, no precipitation was observed during the synthesis. However, slight aggregation of the nanocomposite particles was observed after several days of the reaction. These aggregates could be easily re-dispersed back to a stable colloidal system after sonication. Thus, the PEI and PNIPAm graft chains located in the particle shell were able to provide both electrostatic and steric stabilizations of the resultant nanocomposite particles.

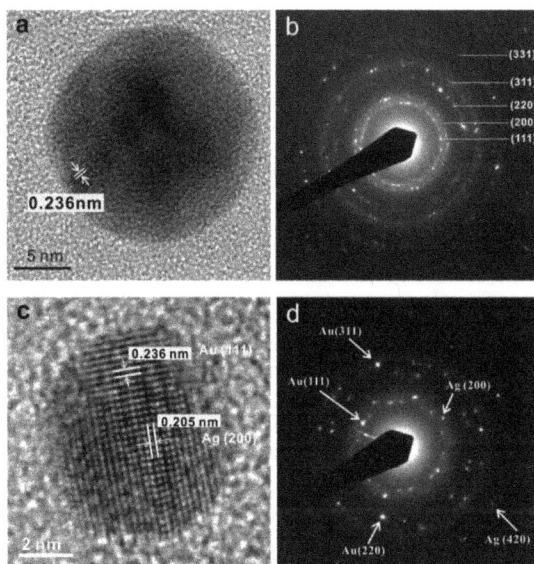

Figure 2. (a) HRTEM image of Au nanoparticle embedded within a microgel template with measured lattice *d*-spacing of 0.236 nm; (b) the selected area electron diffraction (SAED) pattern from several Au nanoparticles; (c) HRTEM image of a magnified single Ag@Au nanocrystal with corresponding lattice parameters of Au and Ag; (d) SAED of Ag@Au nanoparticles.

Table 1. Particle size and *zeta*-potential of different nanocomposite particles.

Type	Mean hydrodynamic diameter	zeta-Potential
	(nm)	(mV)
PNIPAm/PEI	384	+34
Au@PNIPAm/PEI	284	+15
Ag@Au/PNIPAm/PEI	270	+5

3.4. Surface Chemical Composition of Nanocomposite Particles

Surface chemical compositions of both the Au@PNIPAm/PEI and the Ag@Au/PNIPAm/PEI nanocomposite particles were characterized by XPS analysis. This technique detects elements at a depth of 10 nm, and provides surface chemical information based on photoemission of electrons induced by X-rays. Figure 3a shows the XPS spectrum of the Au/PNIPAm/PEI nanocomposite particles. The survey spectrum reveals the characteristic binding energy peaks of C, O, N, and Au. Deconvoluted C 1s, O 1s, and N 1s peaks are shown in Figure S3 in the Supplementary Materials. The deconvoluted C 1s peaks at 285.0 and 287.7 eV are assigned to C–C/C–H and C=O bonds. The deconvoluted O1s peaks at 531.6 and 533.3 eV are assigned to the amide as well as hydroxyl functional groups which come from physically absorbed H_2O. The deconvoluted N 1s peaks at 399.6 and 401.1 eV are assigned to the nitrogen from amine and amide. The XPS elemental peak profile (Figure 3a, inset) of gold nanoparticles is confirmed based on two characteristic peaks at 84.3 and 88.0 eV, which correspond to the Au $4f_{7/2}$ and Au $4f_{5/2}$ of elemental gold at zero oxidation state. These results suggest that the particle shell contains components of PEI, PNIPAM and gold nanoparticles.

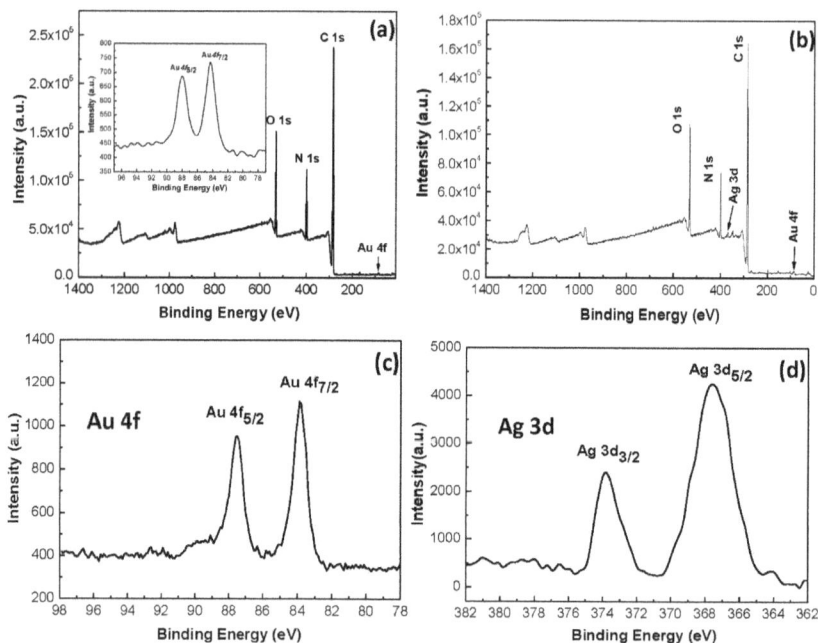

Figure 3. (**a**) Survey scan of Au/PNIPAm/PEI microgel. Inset is Au *4f* XPS spectrum of Au nanoparticles; (**b**) Survey scan of Ag@AuPNIPAm/PEI microgel film (Au/Ag of 50/50 mol. ratio); (**c**) Au *4f* XPS spectrum; and (**d**) Ag *3d* XPS spectrum of Ag@Au bimetallic nanoparticles.

Figure 3b shows the XPS spectrum of the Ag@Au/PNIPAm/PEI nanocomposite particles. The survey spectrum reveals the characteristic binding energy peaks of C, O, N, Au, and Ag. Deconvoluted C 1s, O 1s and N 1s peaks are shown in Figure S4 in the Supplementary Materials. Figure 3c shows two peaks at binding energies of about 83.8 and 87.5 eV, which were assigned to Au $4f_{7/2}$ and Au $4f_{5/2}$ of zero-valent gold (Au^0), respectively. Figure 3d shows the two peaks at binding energies of 367.6 and 373.8 eV, which corresponded to Ag $3d_{5/2}$ and Ag $3d_{3/2}$ of metallic Ag^0, respectively. The differences between the $4f_{7/2}$ and $4f_{5/2}$ peaks for gold nanocrystal (3.6 eV) and between the $3d_{5/2}$ and $3d_{3/2}$ peaks for silver nanocrystal (~6.0 eV) were of similar values of zero valent gold and silver as reported in the literature [30]. These results, in agreement with the results of the HRTEM analysis, suggest that both Au and Ag components were located in the shell region.

3.5. Catalytic Properties of Au and Ag@Au PNIPAm/PEI Nanocomposite Particles

The catalytic activities of the nanocomposite particles were studied through a model reaction to reduce *p*-nitrophenol to *p*-aminophenol with sodium borohydride. The reaction was monitored by a UV-vis spectrometer since *p*-nitrophenol has a characteristic maximum absorption at 400 nm. A pseudo-first order kinetic reaction was chosen to derive the rate of catalytic reaction. At an excess amount of the reducing agent, NaBH$_4$ (mole ratio of NaBH$_4$:*p*-nitrophenol = 6.7:1), the rate of reaction constant, *k* of a pseudo-first order kinetic model can be described by Equation (1):

$$\frac{-dc}{dt} = kC_t \tag{1}$$

where the absorbance of *p*-nitrophenol at *t* = 0 (A_0) and at t (A_t) are proportional to its initial concentration (C_0) and concentration at time t (C_t), respectively. Plotting *ln* (C_t/C_o) *versus* time will give the rate constant based on the slope of *k* (s^{-1}).

In a control experiment, the *p*-nitrophenol reacted only with sodium borohydride to form *p*-aminophenol as the sole product. Changes in peak intensity at 400 nm were monitored in intervals of every two minutes (Supplementary Materials, Figure S5). It was found that there was little change in peak intensity after up to 20 min of reaction, indicating a very slow reduction reaction. The rate constant derived from the slope of the linear curve was $5.4 \times 10^{-3} \cdot s^{-1}$.

When Au@PNIPAm/PEI nanocomposite particles were used, the reaction proceeded approximately 4.5 times faster ($2.44 \times 10^{-2} \cdot s^{-1}$) than without the immobilized AuNPs (Figure 4a). When Ag@Au/PNIPAm/PEI nanocomposite particles were used in the same reaction system, the reaction rate was significantly enhanced, as shown in Figure 4b. The reduction rate constant was $6.20 \times 10^{-1} \cdot s^{-1}$, indicating a much faster reaction rate than the monometallic gold nanoparticle system. Figure 5 shows the corresponding catalytic reaction rates by plotting $\ln(C_t/C_0)$ *versus* reaction time for the reduction of *p*-nitrophenol to *p*-aminophenol. These results indicate that the bimetallic nanocomposite particles gave a reaction 25 times faster than using monometallic gold nanocomposite particles and more than 100 times faster than the reaction without nanocomposite particles. In fact, the catalytic performance of the Ag@Au/PNIPAm/PEI nanocomposite particles was superior to other supported Ag@Au bimetallic nanoparticles reported in the literature. For examples, catalytic rate constants of a metal–organic framework-supported Au@Ag ([31], polystyrene-supported Ag@Au [32], and graphene oxide supported Au-Ag alloy [33] were $4.97 \times 10^{-3} \cdot s^{-1}$, $15.47 \times 10^{-3} \cdot s^{-1}$ and $0.05\ s^{-1}$, respectively. The substantial enhancement in catalytic activity may be attributed to the synergistic effect of the bimetallic nanoparticles derived from their unique electronic and geometrical properties [31,34,35]. It was also suggested that the increase in the number of low coordination number edge site of Ag and corner sites of Au could enhance the catalytic activity [36].

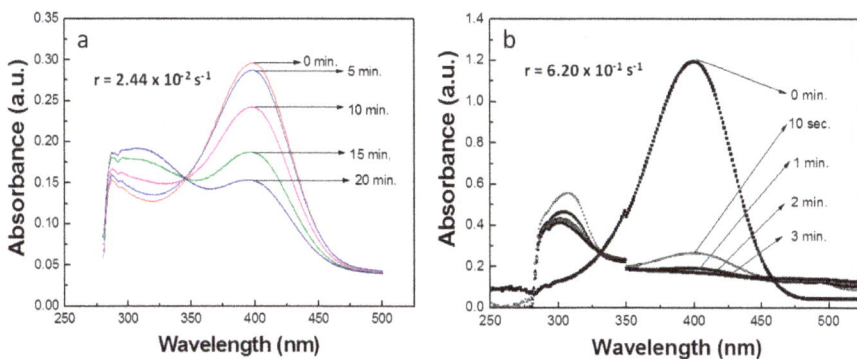

Figure 4. (**a**) Time-dependent UV-vis spectra on the reduction of *p*-nitrophenol to *p*-aminophenol. (**a**) Use of Au@PNIPAm/PEI nanocomposite particles as catalyst; (**b**) use of Ag@Au/PNIPAm/PEI nanocomposite particles as catalyst.

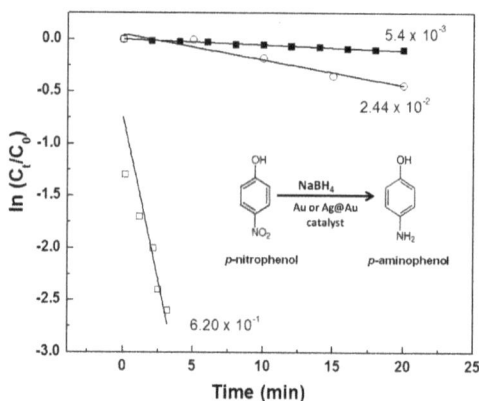

Figure 5. Plots of ln (C_t/C_0) as a function of time for the reaction at room temperature: (■) Without catalyst; (○) In the presence of Au@PNIPAm/PEI (N/Au molar ratio = 28); (□) In the presence of Ag@Au PNIPAm/PEI (N/AgAu molar ratio = 28).

3.6. Stimuli-Responsive Properties of PNIPAm/PEI Template and Tuneable Catalytic Activities of the Nanocomposite Particles

One of the unique properties of the PNIPAm/PEI is its ability to respond to the dual stimuli of pH and temperature. As discussed earlier with reference to the XPS results, there were some PNIPAm chains that co-existed with PEI chains in the particle shell. To verify the morphological changes of the microgels both below and above the volume phase transition temperature (VPTT) of 32 °C, we used atomic force microscopy (AFM) in a fluid mode to observe their morphologies in an aqueous solution. Figure 6a shows an image of the microgel particles at 29 °C with particle sizes in the range of 150 to 200 nm and smooth surface morphology. When the temperature was raised to 45 °C, which is above the VPTT of the microgels, they became not only smaller (100–150 nm), but also had a porous surface form (Figure 6b). The particle size reduction was attributed to the shrinkage of the microgels above their VPTT. The porous surface was generated due to the contraction of the PNIPAm chains located in the shell. When the temperature was cooled down back to 29 °C, the smooth morphology and the original size of the microgel particles were restored. These results suggest that manipulation of solution temperature could induce conformational changes of the microgel with good reversibility. Thus we envisaged that the shrinking and expanding action of the nanocomposite particles could be used to control the catalytic activity of the metallic nanoparticles immobilized within the PNIPAm/PEI microgel through limiting and maximizing the exposure of the metal nanoparticles surface to reactants [37].

The effect of temperature on the catalytic activity of the Au@PNIPAm/PEI nanocomposite particles is shown in Figure 7. Catalytic activity is at the highest value at 25 °C, and decreases as temperature increases, eventually ceasing when the temperature is above 35 °C. The decrease in catalytic activity can be explained based on the accessibility of reactants to gold nanoparticles at different temperatures. When the solution temperature is below the VPTT of the Au@PNIPAm/PEI nanocomposite particles, the template swells and the gold nanoparticles become more accessible under this condition. However, when the temperature is above the VPTT of the nanocomposite particles, the PNIPAm core shrinks notably, resulting in the "dragging in" of the whole microgel particle and the covering of the catalyst surface. As a result, the reactant molecules are more difficult to penetrate and interact with the surface of gold nanocatalyst embedded in the nanocomposite particles for the reaction to proceed. The catalytic activity eventually ceased when most of the surface of the gold nanoparticles was covered by the polymer.

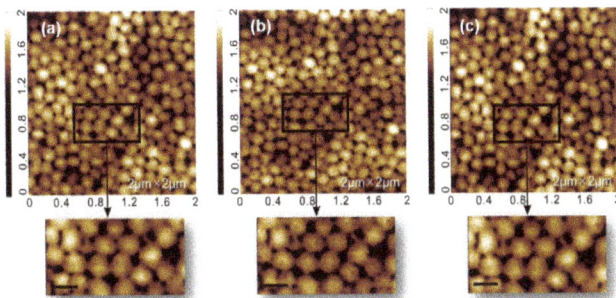

Figure 6. AFM micrographs of PNIPAM/PEI microgel particles measured in a fluid mode at different temperatures: (**a**) 29 °C; (**b**) 45 °C; and (**c**) cooled from 45 to 29 °C. (Scale bar: 200 nm).

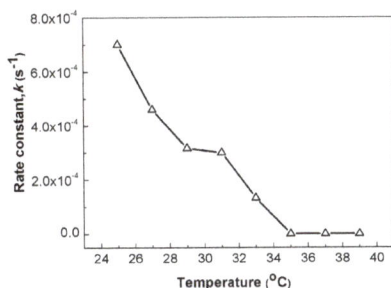

Figure 7. Effect of temperature on the rate constant of Au@PNIPAm/PEI nanocomposite particles for catalytic reduction of *p*-nitrophenol to *p*-aminophenol at pH = 5.6.

The effect of solution pH on the catalytic activity of the Au@PNIPAm/PEI nanocomposite particles was also systematically studied from pH 3 to 11. Results shown in Figure 8 indicate that the highest catalytic activity is at around pH 3 with a corresponding rate constant of $7.5 \times 10^{-3} \cdot s^{-1}$. This activity is almost 10 times higher than at the neutral pH ($7.4 \times 10^{-4} \cdot s^{-1}$). An abrupt reduction in catalytic activity was found in the pH range of 3.0–3.5. Further increasing the solution pH to 7.5 has little influence on the catalytic activity. When increasing the pH to 11, its catalytic activity ceases. To ensure that the acid-catalyzed reaction was minimal or had no effect at all, a control experiment was performed at pH 3 in the absence of nanocomposite particles. The reduction rate of *p*-nitrophenol to *p*-aminophenol at pH 3 was found to be $1.52 \times 10^{-5} \cdot s^{-1}$. This value is much smaller than the reaction using Au/PNIPAm/PEI nanocomposite particles (reaction rate was $7.5 \times 10^{-3} \cdot s^{-1}$) at the same pH of 3. The results confirmed that an enhanced reaction rate at pH 3 was not caused by the acid-catalyzed reaction.

The variation of catalytic activity may be attributed to the pH-responsiveness of the PEI shell of the nanocomposite template. It is known that the percentage of protonated amines varies with the solution pH [38]. For example, percentages of protonated amines at pH 3, 7, and 9 are around 75%, 25%, and 8%, respectively. Therefore, the effect of solution pH on the catalytic activity of the Au@PNIPAm/PEI nanocomposite particles may be explained by the following reasons: Under low acidic pHs (pH = 3–4), a high protonation degree of the amino groups occurs, resulting in the increase of charge density and the stretching of the PEI network. The expanded PEI shell provides more exposure for the gold nanoparticle to interact with reactant molecules. On the other hand, increasing the solution pH leads to lowering the protonation degree, thus forming a more compacted PEI shell. Consequently, the shielding of the PEI shell makes it difficult for the reacting species to diffuse into the catalytic surface of the gold nanoparticles. These results demonstrate that the catalytic activity of Au@PNIPAm/PEI nanocomposite particles can be easily turned "on" and "off" by adjusting the solution pH.

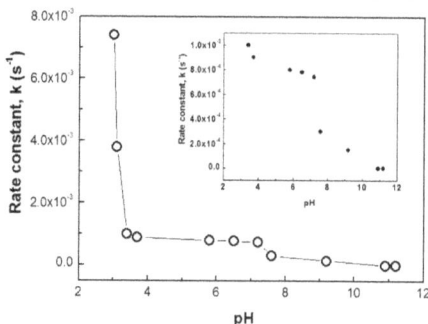

Figure 8. Effect of pH on the rate constant of Au@PNIPAm/PEI nanocomposite particles for catalytic reduction of *p*-nitrophenol to *p*-aminophenol at room temperature.

3.7. Reusability of Nanocomposite Particles

The reusability of Au/PNIPAm/PEI nanocomposite particles has been examined by comparing morphologies of nanocomposite particles before and after the catalytic reactions. The Au/PNIPAm/PEI nanocomposite particles were recovered after one cycle of catalytic reaction through centrifugation and redispersion. TEM images shown in Figure 9 reveal that the morphology of the recovered nanocomposite particles is quite similar to those original nanocomposite particles. These results suggest that the nanocomposite particles may be reusable for catalytic reaction.

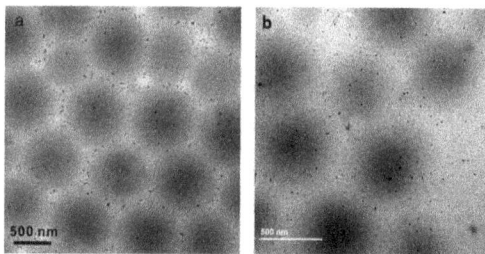

Figure 9. TEM images (**a**) before catalysis and (**b**) after catalysis. The Au/PNIPAm/PEI nanocomposites was recovered at 10,000 rpm centrifugation for 1 hour, followed by redispersing the particles in deionized water.

4. Conclusions

We have developed a simple route for *in situ* synthesis of Au and Ag@Au nanoparticles using smart PNIPAm/PEI core-shell microgel particles as dual reductant and template. The gold nanoparticles were initially formed through a reduction of gold ions with highly concentrated amine functional groups of the microgel. The resultant gold nanoparticles were then used as seeded nanoparticles for further reduction of silver ions to form bimetallic alloy nanoparticles. The use of polymeric amine-based particles to produce metal/polymer nanocomposite particles in an aqueous solution is a simple and green synthesis without using any organic solvent, reducing or stabilizing agents. The resulting nanocomposite particles possess not only excellent catalytic activities with good reproducibility and stability, but also smart properties that allow the tuning of catalytic activities of the

Polymers **2016**, *8*, 105

metal nanoparticles through varying solution pH and temperature. Therefore, these metal nanoparticle immobilized nanocomposite particles possess high potential for practical applications in catalysis.

Supplementary Materials: The supplementary materials can be found at www.mdpi.com/2073-4360/8/4/105/s1.

Acknowledgments: We gratefully acknowledge the Hong Kong Polytechnic University for its financial support of this research and Park Systems, Suwon, Korea for the AFM measurement.

Author Contributions: Noel Peter Bengzon Tan and Pei Li conceived and designed the experiments; Noel Peter Bengzon Tan accomplished the preparation of metal/polymer nanocomposite particles, measurements of particle size, surface charge, and UV spectroscopic study of the catalytic activity of the nanocomposite particles. Cheng Hao Lee performed TEM and XPS analysis; Pei Li, Noel Peter Bengzon Tan, and Cheng Hao Lee analyzed the results and wrote the manuscript.

Conflicts of Interest: The authors declare no conflict of interest.

References

1. Mody, V.V.; Siwale, R.; Singh, A.; Mody, H.R. Introduction to metal nanoparticles. *J. Pharm. Bioall. Sci.* **2010**, *2*, 282–289. [CrossRef] [PubMed]

2. Huang, D.; Yang, G.; Feng, X.; Lai, X.; Zhao, P. Triazole-stabilized gold and related noble metal nanoparticles for 4-nitrophenol reduction. *New J. Chem.* **2015**, *39*, 4685–4694. [CrossRef]

3. Ruíz-Baltazar, A.; Esparza, R.; Rosas, G.; Pérez, R. Effect of the surfactant on the growth and oxidation of iron nanoparticles. *J. Nanomater.* **2015**. [CrossRef]

4. Hu, J.; Yang, Q.; Yang, L.; Zhang, Z.; Su, B.; Bao, Z.; Ren, Q.; Xing, H.; Dai, S. Confining noble metal (Pd, Au, Pt) nanoparticles in surfactant ionic liquids: Active non-mercury catalysts for hydrochlorination of acetylene. *ACS Catal.* **2015**, *5*, 6724–6731. [CrossRef]

5. Murugadoss, A.; Chattopadhyay, A. A "Green" chitosan-silver nanoparticle composite as a heterogeneous as well as micro-heterogeneous catalyst. *Nanotechnology* **2008**, *19*, 15603–15611. [CrossRef] [PubMed]

6. Wang, Z.; Tan, B.; Hussain, I.; Schaeffer, N.; Wyatt, M.F.; Brust, M.; Cooper, A.I. Design of polymeric stabilizers for size-controlled synthesis of monodisperse gold nanoparticles in water. *Langmuir* **2007**, *23*, 885–895. [CrossRef] [PubMed]

7. Bingwa, N.; Meijboom, R. Evaluation of catalytic activity of Ag and Au dendrimer-encapsulated nanoparticles in the reduction of 4-nitrophenol. *J. Mol. Catal.* **2015**, *396*, 1–7. [CrossRef]

8. Gopalan, P.R. Cyclodextrin-stabilized metal nanoparticles: Synthesis and characterization. *Int. J. Nanosci.* **2010**, *9*, 487–494. [CrossRef]

9. Biffis, A.; Orlandi, N.; Corain, B. Microgel-stabilized metal nanoclusters: Size control by microgel nanomorphology. *Adv. Mater.* **2003**, *15*, 1551–1555.

10. Varsha, T.; Namdeo, M.; Mohan, Y.M.; Bajpai, S.K.; Bajpai, M. Review on polymer, hydrogel and microgel metal nanocomposites: A facile nanotechnological approach. *J. Macromol. Sci. Part A* **2008**, *45*, 107–119.

11. Karg, M.; Hellweg, T. New "smart" poly(NIPAM) microgels and nanoparticle microgel hybrids: Properties and advances in characterization. *Curr. Opin. Colloid Interface Sci.* **2009**, *14*, 438–450. [CrossRef]

12. Strong, L.E.; West, J.L. Thermally responsive polymer-nanoparticles composites for biomedical application. *WIREs Nanomed. Nanobiotechnol.* **2011**, *3*, 307–317. [CrossRef] [PubMed]

13. Plaza, H. Antimicrobial polymers with metal nanoparticles. *Int. J. Mol. Sci.* **2015**, *16*, 2099–2116. [CrossRef] [PubMed]

14. Han, D.; Zhang, Q.M.; Serpe, M.J. Poly(*N*-isopropylacrylamide)-*co*-(acrylic acid) microgel/Ag nanoparticles hydrides for the colorimetric sensing of H_2O_2. *Nanoscale* **2015**, *7*, 2784–2789. [CrossRef] [PubMed]

15. Ballauff, M.; Lu, Y. Smart nanoparticles: Preparation, characterization and applications. *Polymer* **2007**, *48*, 1815–1823. [CrossRef]

16. Lu, Y.; Mei, Y.; Drechsler, M.; Baffauff, M. Thermosensitive core-shell particles as carriers for Ag nanoparticles: Modulating the catalytic activity by a phase transition in networks. *Angew. Chem. Int. Ed.* **2006**, *45*, 813–816. [CrossRef] [PubMed]

17. Tan, N.P.B.; Lee, C.H.; Chen, L.; Ho, K.M.; Lu, Y.; Ballauff, M.; Li, P. Facile synthesis of gold/polymer nanocomposite particles using polymeric amine-based particles as dual reductants and templates. *Polymer* **2015**, *76*, 271–279. [CrossRef]

18. Tan, N.P.B.; Lee, C.H.; Li, P. Influence of temperature on the formation and encapsulation of gold nanoparticles using a temperature-sensitive template. *Data Brief.* **2015**, *5*, 434–438. [CrossRef] [PubMed]

19. Sankar, M.; Dimitratos, N.; Miedziak, P.J.; Wells, P.P.; Kiely, C.J.; Hutchings, G.J. Designing bimetallic catalysts for a green and sustainable future. *Chem. Soc. Rev.* **2012**, *41*, 8099–8139. [CrossRef] [PubMed]

20. Park, H.H.; Woo, K.; Ahn, J.P. Core–shell bimetallic nanoparticles robustly fixed on the outermost surface of magnetic silica microspheres. *Sci. Rep.* **2013**, *3*, 1497–1503. [CrossRef] [PubMed]

21. Zhang, X.; Su, Z. Polyelectrolyte-multilayer-supported Au@Ag core-shell nanoparticles with high catalytic activity. *Adv. Mater.* **2012**, *24*, 4574–4577. [CrossRef] [PubMed]

22. Newman, J.D.S.; Blanchard, G.J. Formation of gold nanoparticles using amine reducing agents. *Langmuir* **2006**, *22*, 5882–5887. [CrossRef] [PubMed]

23. Zelewsky, A.V.; Barbosa, L.; Schläpfer, C.W. Poly(ethylenimines) as Brønsted bases and as ligands for metal ions. *Coord. Chem. Rev.* **1993**, *123*, 229–246. [CrossRef]

24. Kobayashi, S.; Hiroshi, K.; Tokunoh, M.; Saegusa, T. Chelating properties of linear and branched poly(ethylenimines). *Macromolecules* **1987**, *20*, 1496–1500. [CrossRef]

25. Herrero, E.; Buller, L.J.; Abruna, H.D. Underpotential deposition at single crystal surfaces of Au, Pt, Ag and other materials. *Chem. Rev.* **2001**, *101*, 1897–1930. [CrossRef] [PubMed]

26. Wang, D.; Li, Y. One-pot protocol for Au-based hybrid magnetic nanostructures via a noble-metal-induced reduction process. *J. Am. Chem. Soc.* **2010**, *132*, 6280–6281. [CrossRef] [PubMed]

27. Cheng, L.C.; Huang, J.H.; Chen, H.M.; Lai, T.C.; Yang, K.Y.; Liu, R.S.; Hsiao, M.; Chen, C.H.; Her, L.J.; Tsai, D.P. Seedless, silver-induced synthesis of star-shaped gold/silver bimetallic nanoparticles as high efficiency photothermal therapy reagent. *J. Mater. Chem.* **2012**, *22*, 2244–2253. [CrossRef]

28. Cuenya, B.R.; Baeck, S.H.; Jaramillo, T.F.; McFarland, E.W. Size- and support-dependent electronic and catalytic properties of Au^0/Au^{3+} nanoparticles synthesized from block copolymer micelles. *J. Am. Chem. Soc.* **2003**, *125*, 12928–12934. [CrossRef] [PubMed]

29. Buffat, P.A.; Flueli, M.; Spycher, R.; Stadelmann, P.; Borel, J.P. Crystallographic structure of small gold particles studied by high-resolution electron microscopy. *Faraday Discuss.* **1991**, *92*, 173–187. [CrossRef]

30. Shanmugam, S.; Viswanathan, B.; Varadarajan, T.K. Photochemically reduced polyoxometalate assisted generation of silver and gold nanoparticles in composite films: A single step route. *Nanoscale Res. Lett.* **2007**, *2*, 175–183. [CrossRef]

31. Jiang, H.L.; Akita, T.; Ishida, T.; Haruta, M.; Xu, Q.J. Synergistic catalysis of Au@Ag core–shell nanoparticles stabilized on metal–organic framework. *J. Am. Chem. Soc.* **2011**, *133*, 1304–1306. [CrossRef] [PubMed]

32. Zhang, S.; Wu, W.; Xiao, X.; Zhou, J.; Xu, J.; Ren, F.; Jiang, C. Polymer-supported bimetallic Ag@AgAu nanocomposites: Synthesis and catalytic properties. *Chem. Asian J.* **2012**, *7*, 1781–1788. [CrossRef] [PubMed]

33. Wu, T.; Ma, J.; Wang, X.; Liu, Y.; Xu, H.; Gao, J.; Wang, W.; Liu, Y.; Yan, J. Graphene oxide supported Au-Ag alloy nanoparticle with different shapes and their high catalytic activities. *Nanotechnology* **2013**, *24*, 125301. [CrossRef] [PubMed]

34. Wang, A.Q.; Liu, J.H.; Lin, S.D.; Lin, T.S.; Mou, C.Y. A novel efficient Au–Ag alloy catalyst system: Preparation, activity, and characterization. *J. Catal.* **2005**, *233*, 186–197. [CrossRef]

35. Benkó, T.; Beck, A.; Frey, K.; Srankóa, D.F.; Geszti, O.; Sáfrán, G.; Marótic, B.; Schay, Z. Bimetallic Ag–Au/SiO2catalysts: Formation, structure and synergistic activity in glucose oxidation. *Appl. Catal. A Gen.* **2014**, *479*, 103–111. [CrossRef]

36. Back, S.; Yeom, M.S.; Jung, Y.S. Active sites of Au and Ag nanoparticle catalysts for CO_2 electroreduction to CO. *ACS Catal.* **2015**, *5*, 5089–5096. [CrossRef]

37. Kawaguch, H. Thermoresponsive microhydrogels: Preparation, properties and applications. *Polym. Int.* **2014**, *63*, 925–932. [CrossRef]

38. Suh, J.; Lee, S.H.; Kim, S.M.; Hah, S.S. Conformational flexibility of poly(ethyleneimine) and its derivative. *Bioorg. Chem.* **1997**, *25*, 221–231. [CrossRef]

polymers

MDPI

Article

Amphiphilic Fluorinated Block Copolymer Synthesized by RAFT Polymerization for Graphene Dispersions

Hyang Moo Lee [1], Suguna Perumal [1,2] and In Woo Cheong [1,2,3,*]

[1] School of Applied Chemical Engineering, Kyungpook National University, 80 Daehakro, Bukgu, Daegu 41566, Korea; hnctk3@naver.com (H.M.L.); suguna.perumal@gmail.com (S.P.)
[2] Research Institute of Advanced Energy, Kyungpook National University, 80 Daehakro, Bukgu, Daegu 41566, Korea
[3] Department of Nano-Science and Technology, Graduate School, Kyungpook National University, 80 Daehakro, Bukgu, Daegu 41566, Korea
* Correspondence: inwoo@knu.ac.kr; Tel.: +82-53-950-7590

Academic Editor: Haruma Kawaguchi
Received: 31 January 2016; Accepted: 16 March 2016; Published: 22 March 2016

Abstract: Despite the superior properties of graphene, the strong π–π interactions among pristine graphenes yielding massive aggregation impede industrial applications. For non-covalent functionalization of highly-ordered pyrolytic graphite (HOPG), poly(2,2,2-trifluoroethyl methacrylate)-*block*-poly(4-vinyl pyridine) (PTFEMA-*b*-PVP) block copolymers were prepared by reversible addition-fragmentation chain transfer (RAFT) polymerization and used as polymeric dispersants in liquid phase exfoliation assisted by ultrasonication. The HOPG graphene concentrations were found to be 0.260–0.385 mg/mL in methanolic graphene dispersions stabilized with 10 wt % (relative to HOPG) PTFEMA-*b*-PVP block copolymers after one week. Raman and atomic force microscopy (AFM) analyses revealed that HOPG could not be completely exfoliated during the sonication. However, on-line turbidity results confirmed that the dispersion stability of HOPG in the presence of the block copolymer lasted for one week and that longer PTFEMA and PVP blocks led to better graphene dispersibility. Force–distance (F–d) analyses of AFM showed that PVP block is a good graphene-philic block while PTFEMA is methanol-philic.

Keywords: graphene; block copolymer; RAFT polymerization; dispersion

1. Introduction

Graphene, a two-dimensional structured material, is a type of carbon allotrope exhibiting a hexagonal structure with sp^2-bonding. This unique structure gives superior properties, such as light transparency [1], mechanical strength [2], thermal conductivity [3,4], and electron mobility [5], amongst others. Owing to these unique and superior properties, many researchers have attempted to incorporate graphene derivatives into nanocomposites [6], solar cell devices [7,8], inkjet printing [9,10], touch panels [11], flexible displays [10], and functional coatings [11]. Nevertheless, graphenes are not readily available due to their poor processability. Recently, several preparation methods for graphene have been developed, such as chemical vapor deposition (CVD) [12], mechanical exfoliation [13,14], molecular assembly [15,16], epitaxial SiC [17], and liquid-phase exfoliation [18,19]. Among these, liquid-phase exfoliation has been considered as the most efficient approach for industrial applications due to its low cost for mass production [20]. Both covalent and non-covalent functionalizations of graphene have been extensively studied in liquid-phase exfoliation. The former includes oxidation of graphite leading to defects (*i.e.*, carboxylic acid, hydroxyl, and carbonyl groups or holes) of graphite structure [20–22]. The latter, on the other hand, can give pristine graphenes, but it requires inclusion, dispersants, or stabilizers in order to avoid a massive aggregation of graphenes after exfoliation [23–25].

In order to prepare stable graphene dispersions with minimal aggregation, various types of solvents (e.g., inorganic, organic, and fluorinated oils) and surfactants (ionic, non-ionic, short, and polymeric) have been exploited [26–29]. A block copolymer dispersant for graphene dispersion has several advantages over short chain surfactants, such as better colloidal stability, minimal use of addition polymeric binders, slow migration, and controllable compatibility. To the best of our knowledge, however, there are only a few reports on block copolymer dispersants for graphene dispersion [30–32]. In this work, amphiphilic block copolymers of poly(2,2,2-trifluoroethyl methacrylate)-*block*-poly(4-vinyl pyridine) (PTFEMA-*b*-PVP) were synthesized by reversible addition-fragmentation chain transfer (RAFT) polymerization while molecular weights, dispersity ($Đ_M$), and compositions of the block copolymers were finely controlled. Graphene dispersions were then prepared from highly-ordered pyrolytic graphite (HOPG) in methanol with four different types of PTFEMA-b-PVP block copolymers, and the characteristics of their dispersion states were investigated.

2. Materials and Methods

2.1. Reagents

2,2,2-Trifluoroethyl methacrylate (TFEMA, Fluorochem, Hadfield, UK) and 4-vinyl pyridine (VP, 95%, Aldrich, St. Louis, MO, USA) were used after purification by an inhibitor remover column (inhibitor removers, Aldrich). 4-Cyano-4-(phenylcarbonothioylthio)pentanoic acid (CTP, 97%, Aldrich), 1,4-dioxane (99.5%, Acros, Geel, Belgium), hexane (95%, Duksan, Ansan, Korea), tetrahydrofuran (THF, 99.5%, Duksan), toluene (anhydrous, 99.8%, Aldrich), chloroform (extra pure grade, Duksan), methanol (99.8%, Duksan), and hydrofluoric acid (50.0%, Duksan) were used as received without further purification. 2,2′-Azobisisobutyronitrile (AIBN, 98%, Junsei, Tokyo, Japan) was used after recrystallization in methanol. HOPG (graphene nanoplatelets, M-25 grade, XG Sciences, Lansing, MI, USA) was used as a source of graphene nanoplatelets. Ultra-pure water (resistivity > 18.2 MΩ·cm, Purelab, Elga, High Wycombe, UK) was used throughout the experiments.

2.2. Synthesis of Poly(2,2,2-trifluoroethyl methacrylate) (PTFEMA) Macro-Reversible Addition-Fragmentation Chain Transfer (RAFT) Agents

RAFT polymerization was carried out to prepare the PTFEMA macro-RAFT agent. The reaction scheme is shown in Figure 1a. AIBN, CTP, and TFEMA with 1:2:100 (or 200) molar ratios were added to a 50 mL round-bottom flask along with an equal amount (g) of 1,4-dioxane as TFEMA, followed by the freeze-pump-thaw cycle three times. Following purging with N_2 gas, the flask was sealed. The reaction mixture was then immersed in a preheated oil bath at 70 °C and stirred for 24 h. The product was purified by precipitation in hexanes three times. The resulting polymer was characterized by Fourier transform infra-red (FT-IR) spectroscopy, nuclear magnetic resonance (^{1}H-NMR), and size exclusion chromatography (SEC).

2.3. Synthesis of Poly(2,2,2-trifluoroethyl methacrylate)-block-poly(4-vinyl pyridine) (PTFEMA-b-PVP) Block Copolymers

PTFEMA-*b*-PVP block copolymers were prepared by RAFT polymerization using PTFEMA macro-RAFT agent. The reaction scheme is shown in Figure 1b. AIBN, PTFEMA macro-RAFT agent, and VP with 1:2:100 (or 400) molar ratios were added to a 25 mL round-bottom flask with an equal amount (g) of 1,4-dioxane as the macro-RAFT agent. The freeze-pump-thaw cycle was repeated for the reaction mixture until the air bubbles disappeared during the cycle and the flask was sealed following an N_2 gas purge. The mixture was then immersed in a preheated oil bath at 70 °C and stirred for 24 h. The resulting PTFEMA-*b*-PVP block copolymer was purified by precipitation in hexane three times.

Following the synthesis procedure outlined above, four different types of block copolymers were obtained as shown in Table 1. The resulting block polymers were characterized by FT-IR and ^{1}H-NMR.

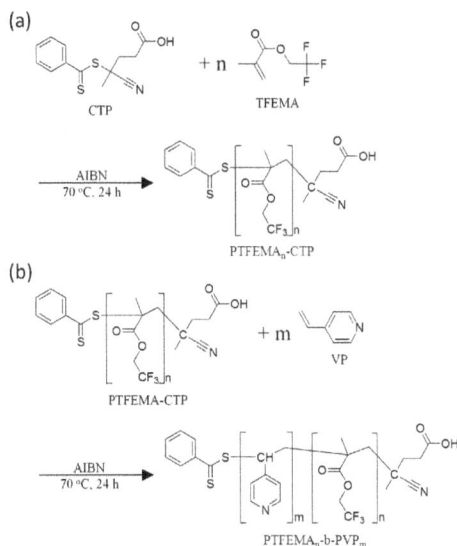

Figure 1. Schematic drawing of the mechanism for syntheses of (**a**) PTFEMA macro-RAFT agent (PTFEMA$_n$-CTP) and (**b**) PTFEMA$_n$-*b*-PVP$_m$ block copolymer. CTP: 4-Cyano-4-(phenylcarbonothioylthio)pentanoic acid; TFEMA: 2,2,2-Trifluoroethyl methacrylate; AIBN: 2,2′-Azobisisobutyronitrile; VP: 4-Vinyl pyridine; PTFEMA-b-PVP: Poly(2,2,2-trifluoroethyl methacrylate)-*block*-poly(4-vinyl pyridine). RAFT: Reversible addition-fragmentation chain transfer.

Table 1. The number average molecular weights and degrees of polymerization of PTFEMA-*b*-PVP block copolymers and the graphene concentrations from the supernatant solution of highly-ordered pyrolytic graphite (HOPG) dispersions with corresponding block copolymers.

Designation	M_n for PTFEMA [1] (g/mol)	M_n for PVP [2] (g/mol)	Degree of polymerization	Graphene concentration [3] (mg/mL)
1	11,092	4,338	PTFEMA$_{66}$-*b*-PVP$_{41}$	0.260
2	11,092	21,581	PTFEMA$_{66}$-*b*-PVP$_{205}$	0.350
3	22,870	3,309	PTFEMA$_{136}$-*b*-PVP$_{31}$	0.275
4	22,870	20,410	PTFEMA$_{136}$-*b*-PVP$_{194}$	0.385

[1] determined by size exclusion chromatography (SEC); [2] determined by proton nuclear magnetic resonance (^{1}H-NMR) peak integration and SEC data of PTFEMA; [3] from the supernatant solution of HOPG methanolic dispersion one week after sonication.

2.4. Preparation of Graphene Dispersions

For industrial purposes, such as conductive ink or barrier coating, it is important to prepare a stable and highly concentrated graphene dispersion. To study the dispersion stability of graphene nanoplatelets, HOPG was dispersed in methanol with PTFEMA-b-PVP block copolymer as a dispersant. For the dispersion, 90.9 mg of HOPG, 9.1 mg of the block copolymer, and 20 mL of methanol were added to a vial. The concentrations of all block copolymers were fixed at 10 wt % of HOPG (0.058 wt % of methanol). The vial was then sonicated in a 40 kHz bath-type sonicator (SD 80H, S-D Ultra Sonic Cleaner, Seoul, Korea) for 2 h.

2.5. Adhesion Force Study Using Modified Atomic Force Microscope (AFM) Cantilever

Adhesion forces between the block (PTFEMA block or PVP block) and graphene surface were investigated by measuring Force–distance (F–d) curves obtained with an atomic force microscope (AFM, XE-7, Park System, Suwon, Korea) with modified cantilevers. To modify the cantilevers for the measurements, a cantilever (NSC-36, Park System) was etched by 2% hydrofluoric acid solution for 1 min to remove the oxide layer on the silicon surface, rinsed by ultra-pure water, dried by an air gun with gentle flowing of N_2 gas, and then immersed into a 50 mL round-bottom flask containing a monomer solution (0.1 M, TFEMA or VP in toluene) at 110 °C. The monomer solution was kept for 2 h, and then the cantilever was rinsed with chloroform. Then, the cantilever was dried with N_2 gas. To prepare the graphene sample, an M-25 powder sample was pelletized by using evacuable pellet dies. F–d curves were taken from 16 points on a graphene surface for one set, with 20 total sets collected. Finally, the average values of adhesion forces were calculated from the measured F–d curves.

2.6. Characterization

To characterize the block copolymers, molecular weights of PTFEMA macro-RAFT agents were measured by SEC (Alliance e2695, Waters, Empower Pro®, Milford, MA, USA). SEC analysis was performed with a refractive index (RI) detector and three different columns (Styragel HR3, Styragel HR4, and Styragel HR5E; Waters, Milford, MA, USA) in series. Tetrahydrofuran (THF, >99.9%, Merck, Darmstadt, Germany) was used as an eluent (35 °C, 1 mL/min). Polystyrene narrow standards (1,060 and 3,580,000 g/mol, Waters; 1,320–2,580,000 g/mol, Shodex, Tokyo, Japan) were used for the calibration. Proton nuclear magnetic resonance (^{1}H-NMR) spectra of PTFEMA macro-RAFT agents and PTFEMA-*b*-PVP block copolymers were obtained using a 500 MHz NMR spectrometer (AVANCE III 500, Bruker, Karlsruhe, Germany) in chloroform-d (with 0.05% TMS, Aldrich, St. Louis, MO, USA). Fourier transform infra-red (FT-IR) spectra were obtained using an FT-IR spectrometer (IR Prestige-21, Shimadzu, Kyoto, Japan) using an ATR attachment (MIRacleA (ZnSe), Shimadzu). The polymers were dissolved in chloroform for the FT-IR measurements.

For dispersion stability, online turbidity data (Turbiscan LAB, Formulaction Co., L'Union, France) and photographic images were taken for a week. After the turbidity analyses, the supernatant graphene solution was isolated for further analyses. The graphene concentration of the supernatant solutions was measured by the gravimetric method with a microbalance (XM 1000P, Sartorius, Göttingen, Germany). For Raman spectroscopy and AFM topography, the isolated supernatant solution was spin coated on a silicon wafer under inert conditions with a spin speed of 700 rpm. In the case of pristine (without block copolymer) HOPG dispersion, there was no graphene in the supernatant solution due to the rapid sedimentation, thus the bottom precipitates were taken for analyses. Raman spectra (Almega X, Thermo scientific, Waltham, MA, USA) at 532 nm wavelength and AFM images were obtained for the precipitate sample. The AFM images were collected using an AFM (XE7, Park system) with a PPP-NCHR (Park system) cantilever in non-contact mode.

3. Results and Discussion

3.1. PTFEMA-*b*-PVP Block Copolymers

In this work, four different block copolymers listed in Table 1 were prepared from stepwise RAFT polymerization. The molecular weights of two PTFEMA macro-RAFT agents were confirmed by SEC and the number average molecular weights (M_n) were 11,092 g/mol ($Đ_M$ = 1.160) and 22,870 g/mol ($Đ_M$ = 1.102), while the degrees of polymerization were calculated as 66 and 136, respectively (refer to Figure 2). The agent with the lesser M_n was used in the second RAFT polymerization for PTFEMA-*b*-PVP block copolymers designated as **1** and **2**, and the agent with the greater M_n was used for **3** and **4**, as shown in Table 1. The molecular weights of PTFEMA-*b*-PVP block copolymers could not be measured using SEC because of the low solubility in typical SEC eluents, e.g., THF, chloroform, and *N,N*-dimethylformamide (DMF). Thus, the molecular weights of the block copolymers were estimated from the proton peak integration of ^{1}H-NMR spectra.

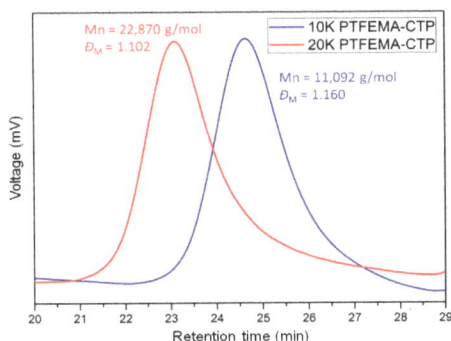

Figure 2. SEC elution curves for two different macro-RAFT agents: PTFEMA$_{66}$-CTP (blue) and PTFEMA$_{136}$-CTP (red).

The degrees of polymerization of PVP block were obtained from the SEC results for PTFEMA macro-RAFT agents and ^{1}H-NMR peak integrals for PTFEMA-*b*-PVP block copolymers. As shown in Figure 3b, peaks a and b (δ 6.39 and 8.33 ppm) are from –ArH protons in PVP block, and peak c (δ 4.34 ppm) is from –CH$_2$CF$_3$ protons in PTFEMA block. From the degree of polymerization for PTFEMA from SEC with the ratios of peak area a, b, and c, the length of PVP block could be calculated. The molecular weights of four PTFEMA-*b*-PVP block copolymers are listed in Table 1. The split peaks –CH$_3$ from PTFEMA block at δ 1.10 and 0.94 ppm are also observed since the protons are hindered.

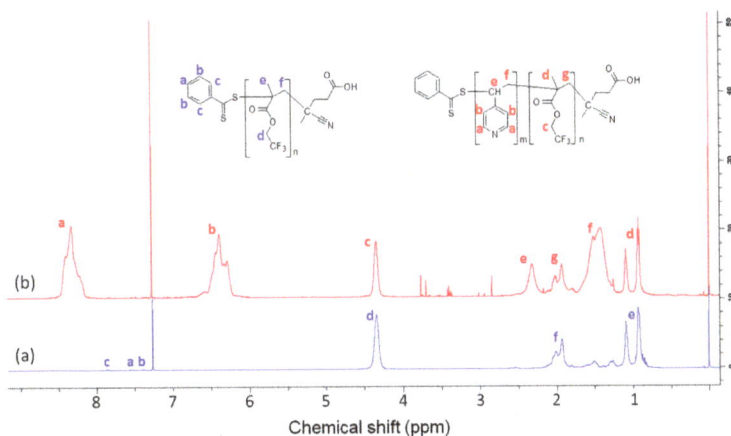

Figure 3. Representative ^{1}H-NMR spectra for (**a**) PTFEMA$_{66}$-CTP and (**b**) PTFEMA$_{66}$-*b*-PVP$_{41}$ block copolymer.

Figure 4 shows the representative FT-IR spectra of PTFEMA$_{66}$-CTP homopolymer and PTFEMA$_{66}$-*b*-PVP$_{41}$ block copolymer. As shown in the FT-IR spectra, a *sp*^{3} –C–H stretching at 3000–2940 cm^{-1}, C=O stretching at 1743 cm^{-1}, C–F stretching at 1278 cm^{-1}, and C–O stretching for ester at 1159 and 1127 cm^{-1} were observed for the TFEMA unit. On the other hand, *sp*^{2} –C–H stretching at 3080–3000 cm^{-1} and aromatic C=C peak at 1598 cm^{-1}, which correspond to the pyridine units of PVP block, are only observed in Figure 4b.

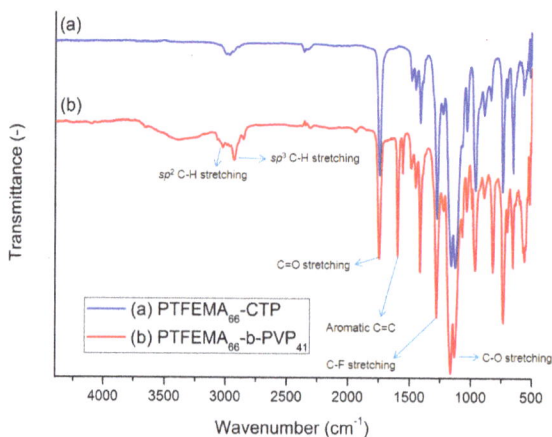

Figure 4. Representative FT-IR spectra for (**a**) PTFEMA$_{66}$-CTP homopolymer and (**b**) PTFEMA$_{66}$-*b*-PVP$_{41}$ block copolymer.

3.2. Adhesion Force Analyses

In order to measure the affinity of each block in PTFEMA-*b*-PVP for the surface of HOPG graphene, F–d curves were recorded using the monomer-modified AFM cantilevers. The adhesion forces were found to be 2.2 ± 0.7 nN for TFEMA and 9.3 ± 0.9 nN for VP, respectively. From the adhesion forces, one can see that VP has 4.2 times greater adhesion force than TFEMA. This suggests that PVP block is more graphene-philic than PTFEMA, despite the greater methanol solubility of PVP relative to PTFEMA. As previously determined, the lone pair electron on the nitrogen atom has an attractive interaction with pi orbitals in a graphene sheet [33]. Due to the strong attractive interaction between the VP and the graphene surface, the PVP block in the block copolymer plays the role of graphene-philic block, while PTFEMA block plays the role of lyophilic block.

3.3. Stability of Graphene Dispersions

Figure 5 shows the dispersion stability of graphene dispersions in terms of turbiscan stability index (TSI), defined as follows:

$$\text{TSI} = \sum_i \frac{\sum_h |\text{scan}_i(h) - \text{scan}_{i-1}(h)|}{H} \tag{1}$$

where H is the length of the sample, h is the height of the measure point, and $\text{scan}_i(h)$ is i^{th} turbiscan intensity at height h. Based on the TSI value, aggregation, precipitation, or creaming of the dispersion can be evaluated. As the dispersion becomes less stable, the TSI value increases [34]. As shown in Figure 5, the time-evolution TSI curve designated as pristine HOPG (M-25, **raw**) surges for the first few hours. Comparably, the TSI curves for the samples designated as **1**, **2**, **3**, and **4** show insignificant change with time—less than one-tenth of the variation seen in sample **raw**—indicating that the block copolymers stabilize graphene in methanol media effectively.

Figure 6 presents photographic images of the graphene dispersions corresponding in time intervals to the TSI analyses in Figure 5. Sample **raw** settled down within 4 h. On the other hand, samples **1**, **2**, **3**, and **4** were stable even for one week. These results confirm that PTFEMA-*b*-PVP block copolymers are suitable for methanolic HOPG graphene dispersions.

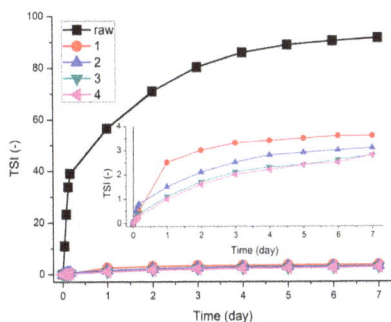

Figure 5. Time-evolution turbiscan stability index (TSI) curves for pristine HOPG (**raw**) and pristine HOPG with PTFEMA$_{66}$-*b*-PVP$_{41}$ (**1**), PTFEMA$_{66}$-*b*-PVP$_{205}$ (**2**), PTFEMA$_{136}$-*b*-PVP$_{31}$ (**3**), and PTFEMA$_{136}$-*b*-PVP$_{194}$ (**4**). The TSI values are calculated for the middle parts of the sample vials shown in Figure 6. The inset shows magnified TSI curves for samples **1–4**.

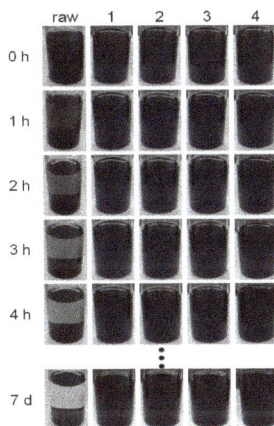

Figure 6. Photographic images for HOPG dispersions *versus* time obtained for pristine HOPG (M-25, **raw**), HOPG with PTFEMA$_{66}$-*b*-PVP$_{41}$ (**1**), PTFEMA$_{66}$-*b*-PVP$_{205}$ (**2**), PTFEMA$_{136}$-*b*-PVP$_{31}$ (**3**), and PTFEMA$_{136}$-*b*-PVP$_{194}$ (**4**).

Further, block copolymers **3** and **4** show little TSI value increases as compared to **1** and **2** (refer to the inset in Figure 5). As listed in Table 1, **2** has a longer PVP block length than **1**, while **3** and **4** have longer PTFEMA blocks than **1** and **2**. As demonstrated in Figure 5, in the case of the samples with the shorter PTFEMA block length (11,092 g/mol), the block copolymer with a longer PVP block length (**2**) shows better stability relative to the copolymer with a shorter PVP block length (**1**). However, for the samples with the longer PTFEMA block length (22,870 g/mol), there was no significant difference in TSI for samples **3** and **4**, despite differences in their PVP block lengths.

It has been known that the destabilization of graphene nanoplatelets in a solution is attributed to strong π–π interactions among graphene surfaces. Aggregation among the graphene nanoplatelets is thermodynamically favorable but the aggregation can be kinetically retarded by using either a stabilizer or a surfactant. PTFEMA-*b*-PVP copolymers can retard the aggregation of graphene nanoplatelets by steric hindrance among adsorbed block copolymers. PTFEMA-*b*-PVP copolymers are anchored at the basal plane of graphene by the adsorption of graphene-philic PVP blocks, while lyophilic PTFEMA blocks are stretched into methanol medium. As shown here, both PTFEMA and PVP block lengths are critical for minimizing the destabilization of graphene.

3.4. Graphene Nanoplatelets

The graphene concentrations of supernatant solutions were measured from the HOPG dispersions by gravimetry one week after the sonication. As shown in Table 1, the graphene concentration was in increasing order as follows: **1 < 3 < 2 < 4**. These results suggest that PVP block length is critical for maintaining stability and preventing graphene nanoplatelets from aggregation. For the preparation of a stable graphene dispersion, both longer lyophilic and graphene-philic blocks are favorable.

In order to study the graphene structure after sonication, Raman and AFM analyses were carried out. In the Raman spectra shown in Figure 7, the D bands are observed at 1360 cm^{-1}. When the hexagonal structure of graphene is disturbed, the intensity of the D band increases. Defectless or perfect graphene usually has an I_D/I_G value of zero, while graphene oxide has an I_D/I_G value of >0.9 [35]. In addition the G band and the 2D band, peaks appear at 1590 and 2700 cm^{-1}, respectively. The thickness of graphene can be estimated from the I_{2D}/I_G value, where the I_{2D}/I_G value for monolayer graphene is >2.0, and the I_{2D}/I_G value for pristine graphite is <0.4 [36–38]. From Figure 7 and Table 2, it is seen that the graphene nanoplatelets are almost like graphite, since the I_{2D}/I_G value for pristine graphite (HOPG, **raw**) is 0.4281. Compared to this, the I_{2D}/I_G values for samples **1**, **2**, **3**, and **4** are slightly increased with the range from 0.4692 to 0.6125. On the other hand, the I_D/I_G value of sample **raw** is 0.14735, but the I_D/I_G values for **1**, **2**, **3**, and **4** are in the range from 0.17071 to 0.26248. The increases of the I_D/I_G for graphene dispersions with block copolymers are due to the increased number of edges in graphene nanoplatelets. Since Raman spectra for **1**, **2**, **3**, and **4** were taken from the supernatant solution of HOPG dispersions, the graphene size would be smaller than in sample **raw**. This small size of graphene has more edge sides working as defects.

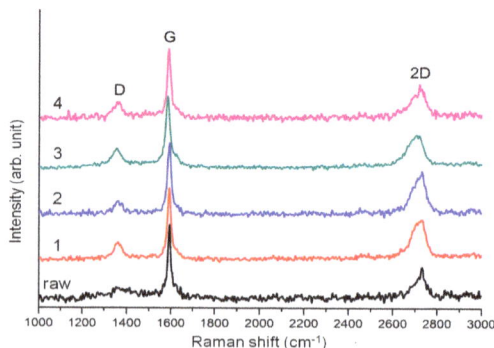

Figure 7. Raman spectra for supernatant solution of HOPG dispersions obtained from pristine HOPG (M-25, **raw**), HOPG with PTFEMA$_{66}$-*b*-PVP$_{41}$ (**1**), PTFEMA$_{66}$-*b*-PVP$_{205}$ (**2**), PTFEMA$_{136}$-*b*-PVP$_{31}$ (**3**), and PTFEMA$_{136}$-*b*-PVP$_{194}$ (**4**).

Table 2. I_{2D}/I_G and I_D/I_G values from Raman spectra for monolayer graphene, graphite, graphene oxide, pristine HOPG (M-25, **raw**), HOPG with PTFEMA$_{66}$-*b*-PVP$_{41}$ (**1**), PTFEMA$_{66}$-*b*-PVP$_{205}$ (**2**), PTFEMA$_{136}$-*b*-PVP$_{31}$ (**3**), and PTFEMA$_{136}$-*b*-PVP$_{194}$ (**4**).

Designation	I_{2D}/I_G	I_D/I_G
Monolayer graphene	>2.0	–
Graphite	<0.4	–
Graphene oxide	–	>0.9
M-25, **raw**	0.4281	0.1474
1	0.5601	0.2250
2	0.6125	0.1707
3	0.4692	0.2625
4	0.4983	0.2307

Figure 8 shows AFM topographic images of the supernatant solutions obtained from graphene dispersion samples. All images show multi-layered graphene nanoplatelets covered with block copolymers. The size distribution of graphene was from a few hundred nanometers to micrometers. The thickness of graphene covered with the polymer aggregates ranged about 6–25 nm, indicating that the number of graphene layers were fewer than 18–73 sheets. Both the graphene surface and the silicon background are covered with the block copolymer aggregates. As for the aggregate size of the block copolymer, sample **4** shows the largest block copolymer aggregates due to the largest molecular weight relative to the other samples.

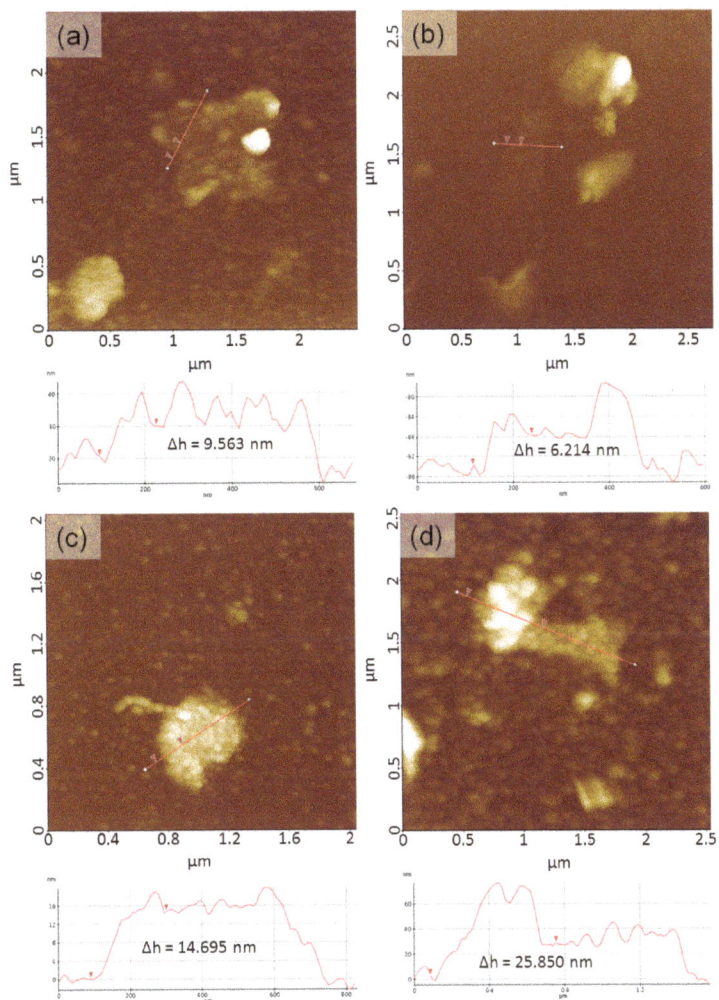

Figure 8. Atomic Force Microscopy (AFM) topographic images of the supernatant solution from HOPG dispersions with (**a**) PTFEMA$_{66}$-*b*-PVP$_{41}$, (**b**) PTFEMA$_{66}$-*b*-PVP$_{205}$, (**c**) PTFEMA$_{136}$-*b*-PVP$_{31}$, and (**d**) PTFEMA$_{136}$-*b*-PVP$_{194}$. Height profiles from the red line on topographic images are depicted below their corresponding images (**a–d**), and the Δh in height profile reveals the height diffrences between red triangles.

Polymers **2016**, *8*, 101

4. Conclusions

Four different types of PTFEMA-*b*-PVP block copolymers were prepared by using stepwise RAFT polymerization. The molecular weights of PTFEMA-CTP homopolymers were measured by SEC, and the molecular weights of PTFEMA-*b*-PVP block copolymers were calculated by SEC and ^{1}H-NMR. Pristine HOPG was dispersed in methanol by using the block copolymers, and the dispersion stability of the resulting graphene dispersions were evaluated using the Turbiscan stability index. Dispersion stability of HOPG in the presence of PTFEMA-*b*-PVP lasted one week and the graphene concentrations of the supernatant solutions ranged from 0.260 to 0.385 mg/mL. Time-dependent backscattering intensity confirmed that TFEMA-*b*-PVP copolymers could substantially retard the aggregation of graphene nanoplatelets. It can be suggested from the F–d curve results that PVP blocks would adsorb at the basal plane of graphene while PTFEMA blocks are soluble in methanol. Therefore, both PVP and PTFEMA block lengths are critical in determining both graphene concentration and dispersion stability.

Acknowledgments: This work was supported by the Ministry of Trade, Industry and Energy, Korea (Grant No. 10044338), Basic Science Research Program through the National Research Foundation of Korea (NRF) funded by the Ministry of Education (Grant No. 2014R1A1A4A01007436), and Korea Agency for Infrastructure Technology Advancement (Grant No. 15CTAP-C077604-02).

Author Contributions: In Woo Cheong conceived and designed the experiments; Hyang Moo Lee prepared and characterized the block polymers and graphene dispersion; Suguna Perumal performed adhesion force study in AFM; Hyang Moo Lee and In Woo Cheong wrote the paper.

Conflicts of Interest: The authors declare no conflict of interest.

References

1. Nair, R.R.; Blake, P.; Grigorenko, A.N.; Novoselov, K.S.; Booth, T.J.; Stauber, T.; Peres, N.M.R.; Geim, A.K. Fine structure constant defines visual transparency of graphene. *Science* **2008**, *320*, 1308. [CrossRef] [PubMed]
2. Lee, C.; Wei, X.; Kysar, J.W.; Hone, J. Measurement of the elastic properties and intrinsic strength of monolayer graphene. *Science* **2008**, *321*, 385–388. [CrossRef] [PubMed]
3. Cai, W.; Moore, A.L.; Zhu, Y.; Li, X.; Chen, S.; Shi, L.; Ruoff, R.S. Thermal transport in suspended and supported monolayer graphene grown by chemical vapor deposition. *Nano Lett.* **2010**, *10*, 1645–1651. [CrossRef] [PubMed]
4. Faugeras, C.; Faugeras, B.; Orlita, M.; Potemski, M.; Nair, R.R.; Geim, A.K. Thermal conductivity of graphene in corbino membrane geometry. *ACS Nano* **2010**, *4*, 1889–1892. [CrossRef] [PubMed]
5. Novoselov, K.S.; Geim, A.K.; Morozov, S.V.; Jiang, D.; Zhang, Y.; Dubonos, S.V.; Grigorieva, I.V.; Firsov, A.A. Electric field effect in atomically thin carbon films. *Science* **2004**, *306*, 666–669. [CrossRef] [PubMed]
6. Kim, H.; Macosko, C.W. Processing-property relationships of polycarbonate/graphene composites. *Polymer* **2009**, *50*, 3797–3809. [CrossRef]
7. Liu, Z.; Liu, Q.; Huang, Y.; Ma, Y.; Yin, S.; Zhang, X.; Sun, W.; Chen, Y. Organic photovoltaic devices based on a novel acceptor material: Graphene. *Adv. Mater.* **2008**, *20*, 3924–3930. [CrossRef]
8. Huang, L.; Huang, Y.; Liang, J.; Wan, X.; Chen, Y. Graphene-based conducting inks for direct inkjet printing of flexible conductive patterns and their applications in electric circuits and chemical sensors. *Nano Res.* **2011**, *4*, 675–684. [CrossRef]
9. Wang, X.; Zhi, L.; Mullen, K. Transparent, conductive graphene electrodes for dye-sensitized solar cells. *Nano Lett.* **2008**, *8*, 323–327. [CrossRef] [PubMed]
10. Shin, K.-Y.; Hong, J.-Y.; Jang, J. Flexible and transparent graphene films as acoustic actuator electrodes using inkjet printing. *Chem. Commun.* **2011**, *47*, 8527–8529. [CrossRef] [PubMed]
11. Bae, S.; Kim, H.; Lee, Y.; Xu, X.; Park, J.-S.; Zheng, Y.; Balakrishnan, J.; Lei, T.; Kim, H.R.; Song, Y.I.; *et al.* Roll-to-roll production of 30-inch graphene films for transparent electrodes. *Nat. Nanotechnol.* **2010**, *50*, 574–578. [CrossRef] [PubMed]
12. Obraztsov, A.N. Chemical vapour deposition: Making graphene on a large scale. *Nat. Nanotechnol.* **2009**, *4*, 212–213. [CrossRef] [PubMed]

13. Chang, Y.M.; Kim, H.; Lee, J.H.; Song, Y.-W. Multilayered graphene efficiently formed by mechanical exfoliation for nonlinear saturable absorbers in fiber mode-locked lasers. *Appl. Phys. Lett.* **2010**, *97*, 211102. [CrossRef]

14. Lin, T.; Chen, J.; Bi, H.; Wan, D.; Huang, F.; Xie, X.; Jiang, M. Facile and economical exfoliation of graphite for mass production of high-quality graphene sheets. *J. Mater. Chem. A* **2013**, *1*, 500–504. [CrossRef]

15. Han, P.; Akagi, K.; Canova, F.F.; Mutoh, H.; Shiraki, S.; Lwaya, K.; Weiss, P.S.; Asao, N.; Hitosugi, T. Bottom-up graphene-nanoribbon fabrication reveals chiral edges and enantioselectivity. *ASC Nano* **2014**, *8*, 9181–9187. [CrossRef] [PubMed]

16. Han, P.; Akagi, K.; Canova, F.F.; Shimizu, R.; Oguchi, H.; Shiraki, S.; Weiss, P.S.; Asao, N.; Hitosugi, T. Self-assembly strategy for fabricating connected graphene nanoribbons. *ACS Nano* **2015**, *9*, 12035–12044. [CrossRef] [PubMed]

17. Norimatsu, W.; Kusunoki, M. Epitaxial graphene on SiC{0001}: Advances and perspectives. *Phys. Chem. Chem. Phys.* **2014**, *16*, 3501–3511. [CrossRef] [PubMed]

18. Ciesielski, A.; Samori, P. Graphene via sonication assisted liquid-phase exfoliation. *Chem. Soc. Rev.* **2014**, *43*, 381–398. [CrossRef] [PubMed]

19. Cui, X.; Zhang, C.; Hao, R.; Hou, Y. Liquid-phase exfoliation, functionalization and applications of graphene. *Nanoscale* **2011**, *3*, 2118–2126. [CrossRef] [PubMed]

20. Li, D.; Müller, M.B.; Jilje, S.; Kaner, R.B.; Wallace, G.G. Processable aqueous dispersions of graphene nanosheets. *Nat. Nanotechnol.* **2008**, *3*, 101–105.

21. Wojtoniszak, M.; Chen, X.; Kalenczuk, R.J.; Wajda, A.; Łapczuk, J.; Kurzewski, M.; Drozkzik, M.; Chu, P.K.; Borowiak-Palen, E. Synthesis, dispersion, and cytocompatibility of graphene oxide and reduced graphene oxide. *Colloids Surf. B Biointerfaces* **2012**, *89*, 79–85. [CrossRef] [PubMed]

22. Pei, S.; Cheng, H.-M. The reduction of graphene oxide. *Carbon* **2012**, *50*, 3210–3228. [CrossRef]

23. Lotya, M.; Hernandez, Y.; King, P.J.; Smith, R.J.; Nicolosi, V.; Karlsson, L.S.; Blighe, F.M.; de, S.; Wang, Z.; McGovern, I.T.; *et al.* Liquid phase production of graphene by exfoliation of graphite in surfactant/water solution. *J. Am. Chem. Soc.* **2009**, *131*, 3611–3620. [CrossRef] [PubMed]

24. Lotya, M.; King, P.; Khan, U.; De, S.; Coleman, J.N. High-concentration, surfactant-stablized graphene dispersion. *ACS Nano* **2010**, *4*, 3155–3162. [CrossRef] [PubMed]

25. Vadukumpully, S.; Paul, J.; Valiyaveettil, S. Cationic surfactant mediated exfoliation of graphite into graphene flakes. *Carbon* **2009**, *47*, 3288–3294. [CrossRef]

26. Hamilton, C.E.; Lomeda, J.R.; Sun, Z.; Tour, J.M.; Barron, A.R. High-yield organic dispersions of unfunctionalized graphene. *Nano Lett.* **2009**, *9*, 3460–3462. [CrossRef] [PubMed]

27. Nuvoli, D.; Valentini, L.; Alzari, V.; Scognamillo, S.; Bon, S.B.; Piccinini, M.; Illescas, J.; Mariani, A. High concentration few-layer graphene sheets obtained by liquid phase exfoliation of graphite in ionic liquid. *J. Mater. Chem.* **2011**, *21*, 3428–3431. [CrossRef]

28. O'Neill, A.; Khan, U.; Nirmalraj, P.N.; Boland, J.; Coleman, J.N. Graphene dispersion and exfoliation in low boiling point solvents. *J. Phys. Chem. C* **2011**, *115*, 5422–5428. [CrossRef]

29. Bourlinos, A.B.; Georgakilas, V.; Zboril, R.; Steriotis, T.A.; Stubos, A.K. Liquid-phase exfoliation of graphite towards solubilized graphenes. *Small* **2009**, *5*, 1841–1845. [CrossRef] [PubMed]

30. Perumal, S.; Park, K.T.; Lee, H.M.; Cheong, I.W. PVP-*b*-PEO block copolymers for stable aqueous and ethanolic graphene dispersions. *J. Colloid Interface Sci.* **2016**, *464*, 25–35. [CrossRef] [PubMed]

31. Zu, S.-Z.; Han, B.-H. Aqueous dispersion of graphene sheets stabilized by pluronic copolymers: Formation of supramolecular hydrogel. *J. Phys. Chem. C* **2009**, *113*, 13651–13657. [CrossRef]

32. Popescu, M.-T.; Tasis, D.; Papadimitriou, K.D.; Gkermpoura, S.; Galiotis, C.; Tsitsilianis, C. Colloidal stabilization of graphene sheets by ionizable amphiphilic block copolymers in various media. *RSC Adv.* **2015**, *5*, 89447–89460. [CrossRef]

33. Voloshina, E.N.; Mollenhauer, D.; Chiappisi, L.; Paulus, B. Theoretical study on the adsorption of pyridine derivatives on graphene. *Chem. Phys. Lett.* **2011**, *510*, 220–223. [CrossRef]

34. Wisniewsk, M.; Szewezuk-Karpisz, K. Removal possibilities of colloidal chromium (III) oxide from water using polyacrylic acid. *Environ. Sci. Pollut. Res.* **2013**, *20*, 3657–3669. [CrossRef] [PubMed]

35. Ullah, K.; Zhu, L.; Ye, S.; Jo, S.-B.; Oh, W.-C. Photocatalytic and reusability studies of novel ZnSe/graphene nanocomposites synthesized via one pot hydrothermal techniques. *Asian J. Chem.* **2014**, *26*, 4097–4102.

36. We, W.; Yu, Q.; Peng, P.; Liu, Z.; Bao, J.; Pei, S.-S. Control of thinkness uniformity and grain size in graphene films for transparent conductive electrodes. *Nanotechnology* **2012**, *23*, 035603.
37. Gayathri, S.; Jayabal, P.; Kottaisamy, M.; Ramakrishnan, V. Synthesis of few layer graphene by direct exfoliation of graphite and a Raman spectroscopic study. *AIP Adv.* **2014**, *4*, 027116. [CrossRef]
38. Das, A.; Chakraborty, B.; Sood, A.K. Raman spectroscopy of graphene on different substrates and influence of defects. *Bull. Mater. Sci.* **2008**, *31*, 579–584. [CrossRef]

polymers

MDPI

Article

Preparation of Uniform-Sized and Dual Stimuli-Responsive Microspheres of Poly(*N*-Isopropylacrylamide)/Poly(Acrylic acid) with Semi-IPN Structure by One-Step Method

En-Ping Lai [1,2], Yu-Xia Wang [2,*], Yi Wei [2] and Guang Li [1]

[1] State Key Laboratory for Modification of Chemical Fibers and Polymer Materials, College of Materials Science and Engineering, Donghua University, Shanghai 201620, China; nemodhu@163.com (E.-P.L.); lig@dhu.edu.cn (G.L.)
[2] National Key Laboratory of Biochemical Engineering, Institute of Process Engineering, Chinese Academy of Sciences, Beijing 100190, China; ywei@ipe.ac.cn
* Correspondence: yxwang@ipe.ac.cn; Tel.: +86-10-8254-5002

Academic Editor: Sebastien Lecommandoux
Received: 3 December 2015; Accepted: 14 March 2016; Published: 17 March 2016

Abstract: A novel strategy was developed to synthesize uniform semi-interpenetrating polymer network (semi-IPN) microspheres by premix membrane emulsification combined with one-step polymerization. Synthesized poly(acrylic acid) (PAAc) polymer chains were added prior to the inner water phase, which contained *N*-isopropylacrylamide (NIPAM) monomer, *N*,*N*′-methylene bisacrylamide (MBA) cross-linker, and ammonium persulfate (APS) initiator. The mixtures were pressed through a microporous membrane to form a uniform water-in-oil emulsion. By crosslinking the NIPAM in a PAAc-containing solution, microspheres with temperature- and pH-responsive properties were fabricated. The semi-IPN structure and morphology of the microspheres were confirmed by Fourier transform infrared spectroscopy (FTIR), scanning electron microscopy (SEM), and transmission electron microscopy (TEM). The average diameter of the obtained microspheres was approximately 6.5 μm, with Span values of less than 1. Stimuli-responsive behaviors of the microspheres were studied by the cloud-point method. The results demonstrated that semi-IPN microspheres could respond independently to both pH and temperature changes. After storing in a PBS solution (pH 7.0) at 4 °C for 6 months, the semi-IPN microspheres remained stable without a change in morphology or particle size. This study demonstrated a promising method for controlling the synthesis of semi-IPN structure microspheres with a uniform size and multiple functionalities.

Keywords: poly(*N*-isopropylacrylamide) (PNIPAM); semi-interpenetrating polymer network (semi-IPN); stimuli-responsive; uniform microspheres; premix membrane emulsification

1. Introduction

Stimuli-responsive polymers have attracted extensive interest due to their special response behavior to external environment changes. Poly(*N*-isopropylacrylamide) (PNIPAM) is one of the prominent thermally reversible polymers that can change its characteristics around the lowest critical solution temperature (LCST, approximately 32 °C in water) [1,2]. PNIPAM microspheres possess a small size and rapid response time and thus have potential use in many high-tech fields including medical diagnostics, controlled drug delivery and enzyme immobilization [3–8]. To broaden its applicability, dual stimuli-responsive PNIPAM-based microspheres have been developed. Among these, the microspheres that respond to both pH and temperature simultaneously are expected to be widely used.

The pH/temperature dual stimuli-responsive behavior of PNIPAM-based microspheres can be obtained in different ways, including the copolymerization with hydrophilic/hydrophobic monomers and the formation of interpenetrating polymer networks (IPNs) [9–12]. However, a disadvantage of the copolymerization method is that the property of each component will interfere with the other [13]. Compared with copolymerization, the formation of IPNs is an effective strategy to overcome this weakness as the individual components show little interference with each other [14]. IPNs are well known as a combination of two polymers in network form with the polymer components entangled physically [15]. Full-IPNs and semi-IPNs are the two main types of IPNs, and the full-IPNs are characterized by the presence of both polymers in the cross-linked state. If one of the components in IPNs is linear and the other is crosslinked, semi-IPNs can be formed. In both semi-IPN and IPN from two-component systems, each phase evolved by the phase separation is forced to form a compatible metastable structure [16].

Generally, PNIPAM-based IPN microgels are prepared by a two-step seed-swelling polymerization. In 2004, Hu and Xia first prepared PNIPAM–PAAc IPN nanogels (approximately 200 nm) using the two-step method [17,18]. Initially, the PNIPAM cross-linked nanogels were synthesized as the seeds for the next polymerization. The PNIPAM nanogels were then immersed in a solution containing AAc monomer and cross-linker, and the IPN structure was formed by *in situ* polymerization of AAc inside the PNIPAM network. The obtained IPN nanogels were found to undergo the volume phase transition at 34 °C, which was similar to the temperature for the PNIPAM nanogels. Subsequently, several studies have been reported on the preparation of PNIPAM-based microgels with IPN or semi-IPN structures. Ma *et al.* utilized a similar two-step method to prepare semi-IPN PNIPAM–PAAc nanocomposite microgels [19]. First, the PNIPAM microgel was prepared via surfactant-free emulsion polymerization using clay as a cross-linker, and then the AAc monomer was polymerized to form PAAc chains within the PNIPAM microgels. The obtained semi-IPN microgels with diameters ranging from 360 to 400 nm could respond independently to both pH and temperature changes. Temperature and pH dual stimuli-responsive hollow nanogels with an IPN structure based on a PAAc network and a PNIPAM network were fabricated by a two-step sequential colloidal template polymerization and the subsequent removal of the cavity templates [20,21]. These microcapsules were approximately 500 nm at room temperature, and possessed pH and temperature stimuli-responsive properties with little mutual interference. Ahmad *et al.* prepared IPN hydrogel microspheres based on PNIPAM and poly(methacrylic acid) (PMAA) by a sequential polymerization method [22]. The prepared microspheres with a diameter of approximately 400 nm exhibited both temperature- and pH-sensitive volume phase transitions.

In summary, the resultant particles mentioned above exhibited improved temperature and pH responsiveness. However, the two-step method was complicated in terms of the production of the IPN or semi-IPN particles. The polymerization time needed to be controlled carefully, otherwise the structures of particles would change from IPN to core–shell structures [17,23]. Additionally, the sizes of the obtained spheres were nano- or sub-micron, which will limit their further applications. For instance, particles are often filled into columns for separation in protein analysis processes, and extremely small spheres will generate high back pressure [24]. Furthermore, it is inconvenient to identify the structure of nano-sized particles in aqueous solution using optical microscopy. Because the properties and behaviors of the microgels are strongly dependent on its water content, a direct inspection of microgels in aqueous solution is very important [25]. Given the above considerations, it would be beneficial to explore an easier strategy for fabricating uniform dual stimuli-responsive micro-sized microspheres with an IPN or semi-IPN structure. Membrane emulsification is a technique that has proven especially useful in preparing uniform-sized particles [26]. According to the preparation mechanisms, this technique is divided into two types: direct membrane emulsification and premix membrane emulsification. Compared with the former, it is more suitable to prepare small particles for the systems with high viscosity in the premix membrane emulsification process. Moreover, the technique characterized by a high trans-membrane flux can obtain uniform droplets efficiently, which would be beneficial for industry production in large scale. Recently, our research group has prepared uniform particles with various structures using this technology [27–29]. Therefore, it is expected that dual stimuli-responsive PNIPAM-based microspheres with semi-IPN structure can be prepared by this novel process.

In this study, we proposed combining the premix membrane emulsification process with the subsequent one-step suspension polymerization method to fabricate dual stimuli-responsive microspheres with a semi-IPN concept. Thus, by carrying out polymerization of NIPAM with cross-linker in PAAc aqueous droplets, the two-component network could be formed directly. This work was focused on the preparation and characterization of semi-IPN structure microspheres, as well as its temperature- and pH-induced volume phase transition behaviors. Furthermore, the stability of microspheres stored in PBS solution was investigated using confocal laser scanning microscopy (CLSM) and a laser particle size analyzer at different times.

2. Experimental Section

2.1. Materials

N-isopropylacrylamide (NIPAM) was purchased from Tokyo Chemical Industrial Co., Ltd. (Tokyo, Japan) and was recrystallized from *n*-hexane at 40 °C before using. Acrylic acid (AAc) and Span 80 were bought from Sinopharm Chemical Regent Beijing Co., Ltd. (Beijing, China), and ammonium persulfate (APS) was supplied by Shantou Xilong Chemical Factory Guangdong, China. *N,N,N′,N′*-tetramethylethylenediamine (TEMED), *N,N′*-methylenebis (acrylamide) (MBA) and Rhodamine 123 (Rh 123) were purchased from Sigma-Aldrich Co., St. Louis, MO, USA. All these reagents of analytical grade were used as received. The micro-porous membrane and membrane emulsification equipment (FM-500M) were kindly provided by Senhui Microsphere Tech (Suzhou, China) Co., Ltd. Deionized water used in the synthesis and characterization was obtained through the RiOs-water system (Millipore Corp., Billerica, MA, USA) to remove impurities.

2.2. Synthesis of PAAc

Six milliliters of acrylic acid, 0.45 g of APS, and 0.5 g of isopropyl alcohol were added into 15 mL of water, after which the mixture was heated to 80 °C. Polymerization was carried out for 2 h in a 100-mL three-neck flask attached to a reflux condenser. The PAAc solution was washed with petroleum ether to remove impurities and then dried at room temperature in a vacuum. Gel permeation chromatography (GPC, Waters Corp., Milford, MA, USA) was used to measure the molecular weight of the polymer using polystyrene as a standard, with 5800 as the number-average molecular weight (M_n) of the PAAc polymer and 1.08 as the M_w/M_n value.

2.3. Preparation of PNIPAM and Semi-IPN Microspheres

The microspheres were synthesized using suspension polymerization (Figure 1), and the recipes of the water phase are shown in Table 1. Generally, different ratios of NIPAM, PAAc, MBA, and APS were dissolved together in 9.5 mL of deionized water. The pH value of the water phase was adjusted to 4.0 using an HCl solution, and the carboxyl groups of PAAc were not ionized as its pKa was 4.7 [30]. In this pH condition, a hydrogen bond could exist between the carboxyl groups of PAAc and the amide groups of NIPAM. Cyclohexane (100 mL) and 2.5 g of Span 80 were mixed together in a beaker. The water phase was then dispersed into the oil phase at 140 rpm for 30 min. The coarse emulsion was pressed through a micro-porous membrane (the pore size of the membrane was 5.2 μm) under a certain nitrogen pressure, and this preliminarily emulsified emulsion was passed through the same membrane in the next pass. The final water-in-oil (W/O) emulsion was obtained after repeating the above process 1–5 times. The obtained emulsion was bubbled with nitrogen gas to remove oxygen before polymerization. Afterwards, TEMED dissolved in 2 mL of cyclohexane was introduced to accelerate the reaction. The system was kept at 20 °C under a nitrogen atmosphere for 4 h.

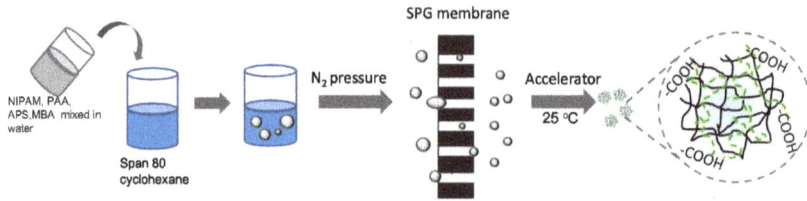

Figure 1. Process for the preparation of PNIPAM/PAAc semi-IPN microspheres.

Table 1. Recipes for the preparation of PNIPAM-based microspheres.

Sample Code	NIPAM ($\times 10^{-1}$ mol/L)	PAAc ($\times 10^{-5}$ mol/L)	MBA ($\times 10^{-2}$ mol/L)	APS ($\times 10^{-2}$ mol/L)	TEMED ($\times 10^{-3}$ mol)
PNIPAM	7.1	0	3.2	1.1	6.7
semi-IPN1	7.1	3.4	3.2	1.1	6.7
semi-IPN2	7.1	3.4	3.2	2.2	13.4
semi-IPN3	7.1	3.4	3.2	3.3	20.1

After the polymerization, the dispersions were washed with acetone and deionized water three times to ensure the complete removal of unreacted chemicals. The obtained microspheres were dispersed in deionized water for further analysis. A series of PNIPAM–PAAc semi-IPN microspheres (semi-IPN1, semi-IPN2, semi-IPN3) were synthesized by changing the initiator APS amount while keeping a constant molar ratio of initiator APS to accelerator TEMED.

2.4. Characterizations of Microspheres

The shape and surface morphology of the microspheres were observed with a scanning electron microscope (SEM, JEOL, JSM-6700F, Tokyo, Japan). Before the measurement, the microspheres were dried by CO_2 supercritical drying with K850 Critical Points Driers (Quorum/Emitech, Ashford, UK) [11]. The resulting samples were coated with gold using a fine coater (JEOL JFC-1600, Tokyo, Japan) on a copper platform.

The morphology of the dried microspheres was examined by transmission electron microscopy (TEM, JEOL, JME-2100, Tokyo, Japan). Before TEM observation, the samples were stained by mixing 1 mL of the microsphere dispersion and 100 μL of a 0.75 mmol·L^{-1} uranyl acetate solution. Subsequently, 15 μL of the mixture was placed on a copper grid (coated with a carbon membrane) and then dried overnight at room temperature.

The particle size and size distribution was measured by laser diffraction using a Mastersizer 2000 (Malvern Instrument, Malvern, UK). The samples were allowed to equilibrate for at least 15 min at each condition. At least three replicates were assessed for each sample to give an average hydrodynamic diameter and size distribution. The particle size distribution was referred to as the Span value and was calculated as follows [31–33]:

$$Span = \frac{D_{V,90\%} - D_{V,10\%}}{D_{V,50\%}},$$

where $D_{V,90\%}$, $D_{V,50\%}$ and $D_{V,10\%}$ are the volume size diameters at 90%, 50% and 10% cumulative volumes, respectively. The smaller the span value indicates the narrower the size distribution.

The chemical structure was analyzed by Fourier transform infrared spectrometer (FTIR, NicoletiS50, Thermo Fisher Scientific Inc., Waltham, MA, USA) using a KBr tablet containing the microsphere powders.

Thermal analysis was performed by differential scanning calorimetry (DSC, TGA/DSC1, Mettler-Toledo Inc., Columbus, OH, USA) in a nitrogen atmosphere. Samples of approximately

4 mg were placed in aluminum sample pans and sealed. The first run was heated from 20 to 200 °C to remove the thermal history, and the second run was heated from 20 to 250 °C at a heating rate of 10 °C/min. An empty aluminum pan of an approximately equal weight was used as a reference.

2.5. Measurement of Thermo- and pH-responsive Behaviors of Microspheres

Thermo- and pH-responsive behaviors of the prepared microspheres were investigated by measuring the transmittance values of the microsphere aqueous dispersions at 575 nm using an UV-vis spectrometer. Microsphere dispersions were treated in an ultrasound instrument for 15 min before the measurements. During the thermo-responsive behaviors measurement, the temperatures were varied from 25 to 50 °C using a circulating water pump, and the heat rate was approximately 0.1–0.25 °C/min. The LCST value was judged to be the initial break point of the curve of transmittance *versus* temperature. For pH-responsive behavior measurements, a 0.1 M HCl solution and a 0.1 M NaOH solution were used to adjust the pH values of the dispersions. Every temperature or pH point was maintained for 10 min before reading the transmittance, and the measurements were repeated three times.

2.6. Examination of Microsphere Stability

Confocal laser scanning microscopy (CLSM, TCS SP5, Leica Microsystems, Wetzlar, Germany) was used to observe the microspheres in PBS solution (10 mM, pH 7.0), and the samples were labeled by Rhodamine 123 (Rh 123). At predetermined times, a small amount of microspheres (8 mg/mL) was removed from the samples and washed by water to remove the free PAAc chains. The microspheres were then added into the Rhodamine 123 solution (5 μg/mL) and incubated at 4 °C for 24 h. The labeled microspheres were separated from the unbound dye by centrifugation and were observed using CLSM at an excitation wavelength of 488 nm. For accuracy in the experiments, the same amount of microspheres, dye, and fluorescence intensity were maintained for the different samples.

3. Results and Discussion

3.1. Optimization of the Process Parameters for Preparing Microspheres with a Narrow Size Distribution

3.1.1. Trans-Membrane Pressure

In the premix membrane emulsification process, the preparation of uniform emulsion droplets is based on extruding the primary coarse emulsions through a membrane by a suitable pressure. Thus, the trans-membrane pressure is one of the crucial parameters influencing the size distribution. To investigate the effect of the pressure on particle size, the primary coarse emulsions of semi-IPN3 were pressed through a micro-porous membrane under different trans-membrane pressures, *i.e.*, 235, 255, and 275 kPa. The number of trans-membrane passes used was three, and the other parameters were the same as mentioned in Section 2.3. The relationship between the trans-membrane pressure and the size distribution of the microspheres is shown in Figure 2. It can been observed that the Span values of the resultant microspheres (3.462, 0.266, and 1.081 for 235, 255 and 275 kPa, respectively) were significantly influenced by pressure, and the narrowest size distribution of the microspheres was obtained at 255 kPa. Because they were at a lower pressure, the droplets of coarse emulsions were difficult to break and would pass through the membrane pores by changing their shape. In contrast, a higher pressure caused a faster pass-through rate and promoted the coarse emulsions to cross the membrane at a high speed. In this situation, a large amount of smaller droplets was obtained. Therefore, too low or too high of a trans-membrane pressure could lead to a broad size distribution of the microspheres, and 255 kPa was chosen as the suitable transmembrane pressure to prepare uniform semi-IPN microspheres in the following experiments.

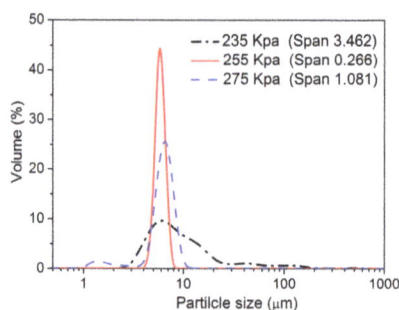

Figure 2. Size distributions of the semi-IPN3 microspheres prepared under different trans-membrane pressures.

3.1.2. Number of Trans-Membrane Passes

The number of trans-membrane passes is another key factor influencing the uniformity of microspheres. The primary coarse emulsions of semi-IPN3 were pressed through the micro-porous membrane for a different number of trans-membrane passes, and the other conditions were the same as mentioned in Section 2.3. As shown in Figure 3, a large amount of larger particles was obtained when the coarse emulsion was passed through the membrane only one time. The reason for this was that the droplets could not be fully emulsified by passing through the membrane pores only once. When the coarse emulsion was pressed through the membrane three and five times, the larger droplets were broken into smaller ones, and a narrower size distribution was obtained. The size distributions of the microspheres prepared by three and five trans-membrane passes were almost the same. For simplicity of operation, three passes were employed in the following experiments. Similar results have been reported in other studies [28,34].

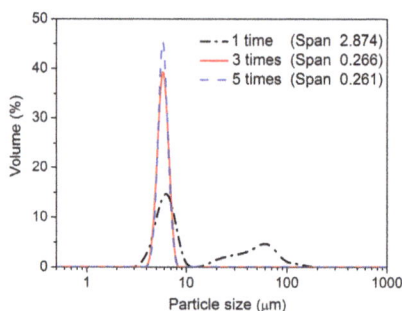

Figure 3. Size distributions of the semi-IPN3 microspheres prepared under different number of trans-membrane passes.

3.2. Chemical Structure of the PNIPAM-PAAc Semi-IPN Microspheres

FTIR spectroscopy was carried out to analyze the chemical compositions and molecular structures of the PNIPAM-PAAc semi-IPN microspheres. As depicted in Figure 4, the typical double peaks at 1650 and 1540 cm^{-1} could be assigned to the amide I band (C=O stretching) and the amide II band (N–H in-plane bending vibration) of PNIPAM, respectively. A band at 1720 cm^{-1} attributed to the stretching vibration of the carboxyl group (–COOH) could be found in the spectrum of PAAc. The spectra of semi-IPN3 showed the characteristic amide groups at 1650 and 1540 cm^{-1} of PNIPAM, and the carbonyl stretching bonds shifted to 1730 cm^{-1} of PAAc. A shift of wavenumbers to higher values on the carboxyl group in the semi-IPN microspheres was consistent with previously published reports [35,36]. These results were the consequence of the hydrogen bond between the carboxyl group of the PAAc chains and the amide group of the PNIPAM network, indicating that PNIPAM–PAAc semi-IPN microspheres were prepared successfully.

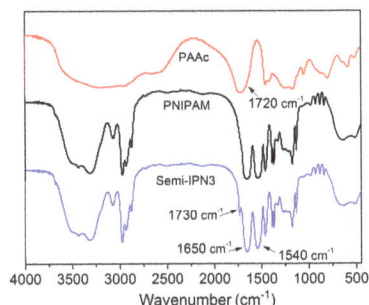

Figure 4. FT-IR spectra for the PAAc, PNIPAM, and semi-IPN3 microspheres.

3.3. Particle Morphology and Size of the PNIPAM-PAAc Semi-IPN Microspheres

To observe the morphology of the microspheres in solid state, SEM measurements were performed. As shown in Figure 5, the surface of the PNIPAM microspheres were largely deformed compared with those of the PNIPAM–PAAc semi-IPN microspheres, which was due to the uneven collapse of the structure as a result of drying. These figures clearly show that the semi-IPN microspheres were more rigid than the PNIPAM microspheres. Some researchers have reported that the introduction of the semi-IPN structure and linear polymer chains could reinforce the unvarnished hydrogel network [37,38]. Thus, in the present system, it could be considered that the presence of the PAAc chains supported the PNIPAM network, resulting in less shrinkage during the dehydration process. Moreover, the surface of the semi-IPN3 microspheres was smoother than that of semi-IPN1 and semi-IPN2. The reason for this was that the increasing concentration of initiator led to a higher crosslinking density, increasing the rigidity of the microspheres [39].

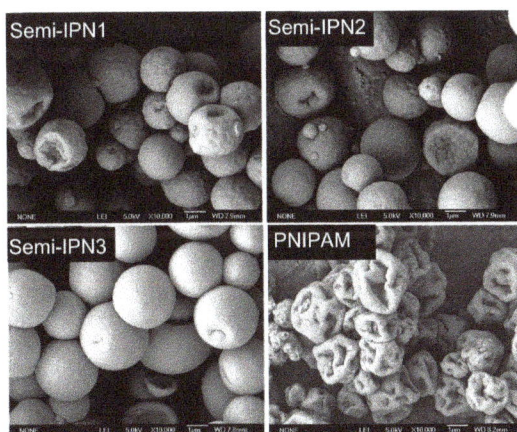

Figure 5. SEM images of semi-IPN and PNIPAM microspheres.

Figure 6 shows the particle size distributions of different microspheres, and Table 2 displays the particle sizes and the yield of the microspheres. It can be observed that the particle sizes of the semi-IPN microspheres (6.7, 6.3 and 6.2 µm, respectively) were similar to those of the PNIPAM microspheres (6.5 µm). However, the Span values of all the semi-IPN microspheres were smaller than those of the PNIPAM microspheres (1.571 ± 0.500), which meant that the size distribution of the semi-IPN microspheres was narrower than that of the PNIPAM microspheres. It has been reported that Span values of less than 1 were considered monosized distributions [40,41]. Thus, the uniform semi-IPN microspheres were obtained, and the average yield of the semi-IPN microspheres was 73%. Remarkably, the particle size distribution of semi-IPN1 was slightly different from those of semi-IPN2 and semi-IPN3, possibly because when the initiator concentration was 1.1×10^{-2} mol/L, the cross-linking density was smaller and the rigidity of the semi-IPN1 microspheres was lower. Hence, a tiny proportion of the microspheres could have been broken during the wash process, which increased the Span value.

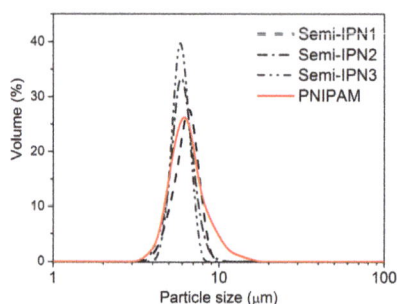

Figure 6. Particle size distribution of the semi-IPN and PNIPAM microspheres (10 mM, pH 7.0 PBS, 25 °C).

Table 2. Particle sizes and Span values of the different microspheres (10 mM, pH 7.0 PBS, 25 °C).

Sample	Particle Size (µm)	Span
PNIPAM	6.5 ± 1.7	1.571 ± 0.500
Semi-IPN1	6.7 ± 2.5	0.649 ± 0.219
Semi-IPN2	6.3 ± 1.1	0.272 ± 0.132
Semi-IPN3	6.2 ± 0.7	0.266 ± 0.112

3.4. TEM Observation of PNIPAM-PAAc Semi-IPN Microspheres

Transmission electron microscopy (TEM) was used for a further in-depth investigation of the microstructure of the PNIPAM–PAAc semi-IPN microspheres. Uranyl acetate was chosen to stain the microspheres before the measurement. Because PNIPAM cannot be stained, the black spots in the sample indicate the location of the PAAc domains. As shown in Figure 7, the amount of large-size black spots in the microspheres decreased from semi-IPN1 to semi-IPN3, and the amount of small-size black spots increased; moreover, a homogeneous texture associated with a network-like feature appeared in the semi-IPN3. These TEM pictures revealed that the increased initiator amount resulted in smaller PAAc and PNIPAM domains.

3.5. DSC Thermograms of the PNIPAM–PAAc Semi-IPN Microspheres

The glass transition temperature (T_g) is a very important characteristic parameter of a polymer as it determines the miscibility of the polymer component. A DSC study was performed on the dried microspheres to obtain the T_g during the heating process, and the calorimetric endotherms are displayed in Figure 8. The T_g of PNIPAM and PAAc were 147 and 92 °C, respectively. PNIPAM–PAAc semi-IPN3 exhibited two T_g values of 97 and 145 °C, which was indicative of a phase-separated structure. Compared with pristine PNIPAM and PAAc microspheres, the T_g of PAAc and PNIPAM in the semi-IPN3 microspheres shifted towards each other. It has been reported that most IPNs form immiscible compositions during synthesis [42]. In addition, it is known from the literature that the T_g of the individual components shift towards each other, indicating a partial mixing of the networks [43,44]. Therefore, the DSC results of the present work showed that an incomplete compatible state existed between the two polymer networks. Meanwhile, the PAAc chains were entangled with the PNIPAM networks and were relatively independent of the PNIPAM networks.

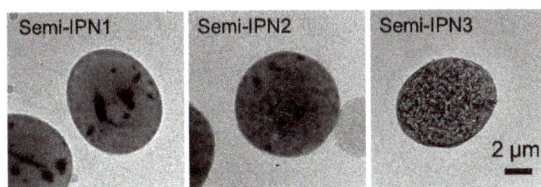

Figure 7. TEM images of semi-IPN with different initiator amounts (samples were labeled with uranyl acetate).

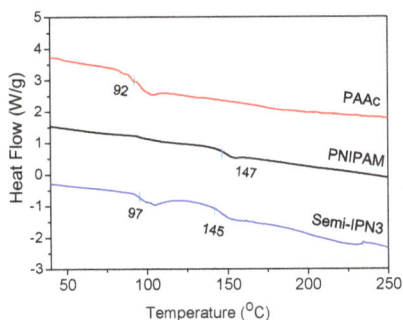

Figure 8. DSC curves of the PAAc, PNIPAM, and semi-IPN3 microspheres.

3.6. Thermo- and pH-Responsive Behaviors of Microspheres

The cloud-point method was used to observe the real-time change in the transmittance of microsphere suspensions in different temperature or pH conditions. The onset temperatures determined by transmittance *versus* the temperature curves, which corresponded to the first sign of microsphere aggregation in the solution, could be equal to the LCST [45]. Figure 9 shows the transmittance changes in the microsphere suspensions at temperatures ranging from 25 to 50 °C. For all semi-IPN samples, the transmittance values decreased significantly at temperatures approximately 32 °C, which was similar to the observations of the PNIPAM microspheres. Other researchers have also measured the transmittances of the suspensions of PNIPAM-*co*-AAc copolymer microspheres at various temperatures, and the results demonstrated that the AAc component could significantly increase the LCST value of the matrix [46]. However, in our experiment, an increase in the LCST values for the semi-IPNs was not observed. This result suggested that due to the chemical independence, the PAAc chains barely hindered the phase transition behavior of the PNIPAM network.

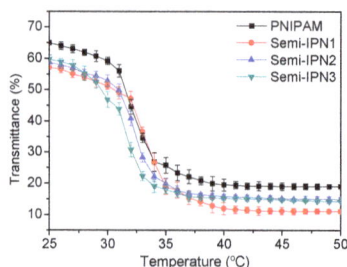

Figure 9. Transmittance of the PNIPAM and semi-IPN microsphere suspensions in different temperatures at pH 7.0.

To investigate the effect of pH on the swelling capacity of PNIPAM and semi-IPN microspheres, the microspheres were immersed in a buffer solution with a pH adjusted from 3.5 to 9.5. The transmittances of the microsphere suspensions as a function of pH are shown in Figure 10 and Table S1 (Supplementary Materials). No transmittance variation in the PNIPAM microsphere suspension was observed in the measured range, whereas the semi-IPN ones showed different transmittance values depending on the pH of the medium. When considered in detail, the curves in Figure 10 could be divided into two sections by pH 5.5. As the pH increased from 3.5 to 5.5, the transmittance increased obviously, indicating an increase in the particle diameter. Indeed, the particle size of microspheres at the various pH values was also measured, which showed similar results (Table S2 in Supplementary Materials). The reasons for this are as follows. When the pH was lower than the pKa of the PAAc component, due to the hydrogen bonding interactions between the amide group of PNIPAM and the carboxyl group of PAAc, the semi-IPN microspheres were in a shrunken state. As the pH increases above the pKa, the carboxyl groups were gradually dissociated, leading to a swelling of the microspheres derived from the increased osmotic pressure and electrostatic repulsion between the charged groups [47]. The transmittance increased with the pH until a plateau region was reached around pH 5.5, with similar behavior reported for PNIPAM–PAAc semi-IPN nanogels with inorganic clay as a crosslinker [19]. These results could be attributed to the completed disassociation of the carboxyl group at a pH higher than the pKa [48].The thermo-sensitive behaviors of the pristine PNIPAM and semi-IPN microspheres were also investigated at different pH values (Figure S1 in Supplementary Materials). It could be observed that as the pH increased from 4.5 to 8.5, the LCST values of the semi-IPN microspheres were increased slightly, whereas those of the PNIPAM microspheres were constant (Table S3 in Supplementary Materials). The results showed that the thermo-sensitive behaviors of the semi-IPN microspheres were barely influenced by the pH value and that the temperature-responsive and pH-responsive behaviors of the semi-IPN microspheres had little effect on each other.

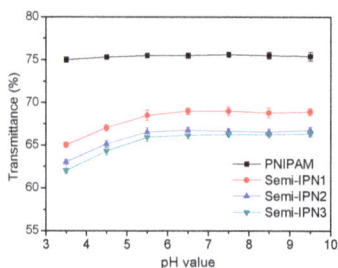

Figure 10. Transmittance of the PNIPAM and semi-IPN microsphere suspensions in different pH conditions at 25 °C.

3.7. Storage Stability of Microspheres

Unlike covalently cross-linked structures, physical entanglement is usually not permanent [49]. In the semi-IPNs system, the PAAc chains and PNIPAM network interacted with each other by hydrogen bond and physical entanglement; thus, it was possible for a fraction of the PAAc chains to diffuse out of semi-IPN microspheres during the storage period. In the enzyme immobilization process, the medium is always a pH 7.0 PBS solution, and the microspheres are usually kept in this solution. The change of functional groups in the microspheres greatly influenced the morphology and the dispersibility of the microspheres, and thus would change the configuration of the enzyme molecules. Taking this into account, an investigation of the morphology and particle size was used to study the stability and dispersibility of microspheres. As shown in Figure 11, all the semi-IPN microspheres were labeled well as Rhodamine 123 (Rh123) has amino groups that would interact with the carboxyl groups of the microspheres. After three months, no obvious change in the fluorescence distribution of the samples was observed. Because only PAAc chains were labeled by Rh123 in the microspheres, the leaching phenomenon of PAAc was hardly observed, which implied that the PAAc chains stably existed in the semi-IPN microspheres over the three months. With a storage time of up to six months, all the samples still had good dispersibility. The morphologies of the semi-IPN2 and semi-IPN3 microspheres were almost the same as the original, whereas there was a little change in the fluorescence distribution in the semi-IPN1. The fluorescence tended to distribute near the surface of the semi-IPN1 microspheres, suggesting that the PAAc chains in the semi-IPN1 could diffuse out of the microspheres easier than the other samples. Combined with the TEM images, the reason for this might be a loose structure in the semi-IPN1.

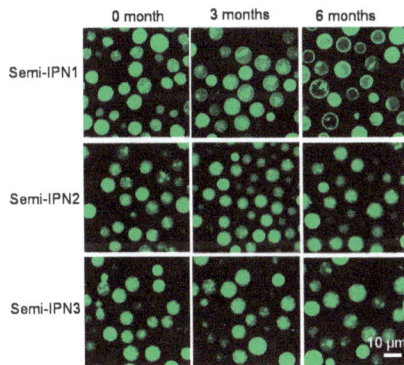

Figure 11. CLSM images of the semi-IPN microspheres at different storage times (4 °C, 10 mM pH 7.0 PBS solution; the samples were labeled with Rh 123).

Figure 12 presents the particle size distribution of the PNIPAM and semi-IPN microspheres at different storage times. It was observed that both the PNIPAM and the semi-IPN microspheres had a narrow size distribution at the initial time. After three months, a bigger size distribution of the PNIPAM microspheres started to emerge. Six months later, the original size distribution of the PNIPAM microspheres disappeared and a bimodal size distribution appeared. The bigger size distribution indicated that the PNIPAM appeared to form aggregates over a prolonged time. The generation of PNIPAM microsphere aggregation might be derived from the absence of charged groups. In contrast, the semi-IPN microspheres retained a narrow size distribution during the whole measurement period. These results, combined with the CLSM images, indicated that the motility of the PAAc chains were restricted by the PNIPAM network in the semi-IPN microspheres, and the PAAc chains in turn improved the dispersibility of the microspheres. In general, the particle size and morphology of the semi-IPN microspheres were relatively stable compared with the PNIPAM microspheres over several months, which indicated that the storage stabilities of the semi-IPN microspheres were more desirable than the PNIPAM microspheres.

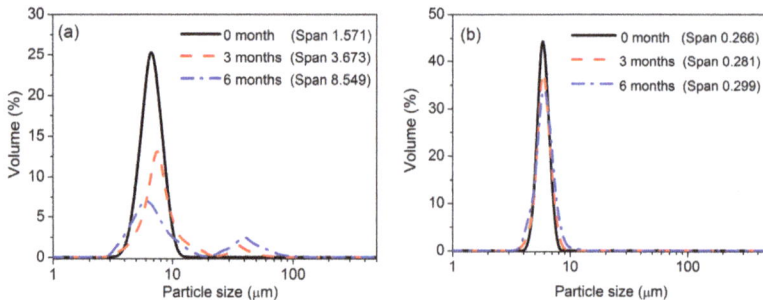

Figure 12. Size distribution of the PNIPAM (**a**) and semi-IPN3 (**b**) microspheres at different storage times (4 °C, 10 mM pH 7.0 PBS solution).

4. Conclusions

In this work, we developed a novel one-step method to synthesize uniform dual stimuli-responsive microspheres with semi-IPN structures. By polymerization of NIPAM monomers with an MBA cross-linkers in an AAc aqueous solution, PNIPAM–PAAc semi IPN microspheres were formed directly. The morphologies of the dried microspheres were observed by SEM. The results revealed that the introduction of the semi-IPN structure and the PAAc chains could reinforce the rigidity of the hydrogel network, and uniform and spherical semi-IPN microspheres were obtained. Both the PNIPAM and semi-IPN microspheres possessed a particle size of approximately 6 μm, and the particle size distribution of the latter was narrower than the former. The semi-IPN microspheres underwent a similar sharp volume phase transition at 32 °C, which was barely influenced by the interpenetration of hydrophilic PAAc. The particle sizes of the semi-IPN microspheres changed with increasing pH values, which was due to the contribution of the PAAc. In addition, the temperature-responsive and pH-responsive properties had little interference with each other. After storage in a PBS solution at 4 °C for 6 months, the morphology and particle size of the semi-IPN microspheres were more stable than those of the PNIPAM microspheres. These studies on network formation and properties provide a reference for designing dual-responsive micro-size hydrogel particles and microspheres with a uniform size, which have various advantages for use in different applications, such as use as drug carriers and enzyme support.

Supplementary Materials: Supplementary Materials can be found at www.mdpi.com/2073-4360/8/3/90/s1.

Acknowledgments: We would like to thank Guanghui Ma for her invaluable help and contribution of theoretical guidance for this article. This study was supported by the Beijing Municipal Science and Technology Commission (Z141100000214007) and the National Natural Science Foundation of China (contract nos. 51173187, 21336010)

Author Contributions: En-Ping Lai performed the experiments and data analysis and wrote the paper. Yu-Xia Wang managed the project as a principal investigator and designed the experiments. Guang Li and Yi Wei contributed to the editing of the paper.

Conflicts of Interest: The authors declare no conflict of interest.

References

1. Gil, E.; Hudson, S. Stimuli-reponsive polymers and their bioconjugates. *Prog. Polym. Sci.* **2004**, *29*, 1173–1222. [CrossRef]
2. Ono, Y.; Shikata, T. Hydration and dynamic behavior of poly(*N*-isopropylacrylamide)s in aqueous solution. *J. Am. Chem. Soc.* **2006**, *128*, 10030–10031. [CrossRef] [PubMed]

3. Pinheiro, J.P.; Moura, L.; Fokkink, R.; Farinha, J.P. Preparation and characterization of low dispersity anionic multiresponsive core-shell polymer nanoparticles. *Langmuir* **2012**, *28*, 5802–5809. [CrossRef] [PubMed]
4. Hendrickson, G.R.; Smith, M.H.; South, A.B.; Lyon, L.A. Design of multiresponsive hydrogel particles and assemblies. *Adv. Funct. Mater.* **2010**, *20*, 1697–1712. [CrossRef]
5. Thorne, J.B.; Vine, G.J.; Snowden, M.J. Microgel applications and commercial considerations. *Colloid. Polym. Sci.* **2011**, *289*, 625–646. [CrossRef]
6. Welsch, N.; Becker, A.L.; Dzubiella, J.; Ballauff, M. Core–shell microgels as "smart" carriers for enzymes. *Soft Matter* **2012**, *8*, 1428–1436. [CrossRef]
7. Wu, Q.; Su, T.; Mao, Y.; Wang, Q. Thermal responsive microgels as recyclable carriers to immobilize active proteins with enhanced nonaqueous biocatalytic performance. *Chem. Commun.* **2013**, *49*, 11299–11301. [CrossRef] [PubMed]
8. Kawaguchi, H. Thermoresponsive microhydrogels: Preparation, properties and applications. *Polym. Int.* **2014**, *63*, 925–932. [CrossRef]
9. Kratz, K.; Hellweg, T.; Eimer, W. Influence of charge density on the swelling of colloidal poly(*N*-isopropylacrylamide-*co*-acrylic acid) microgels. *Colloids Surf. A* **2000**, *170*, 137–149. [CrossRef]
10. Burmistrova, A.; Richter, M.; Eisele, M.; Üzüm, C.; von Klitzing, R. The effect of co-monomer content on the swelling/shrinking and mechanical behaviour of individually adsorbed PNIPAM microgel particles. *Polymers* **2011**, *3*, 1575–1590. [CrossRef]
11. Si, T.; Wang, Y.; Wei, W.; Lv, P.; Ma, G.; Su, Z. Effect of acrylic acid weight percentage on the pore size in poly(*N*-isopropylacrylamide-*co*-acrylic acid) microspheres. *React. Funct. Polym.* **2011**, *71*, 728–735. [CrossRef]
12. Li, Z.; Shen, J.; Ma, H.; Lu, X.; Shi, M.; Li, N.; Ye, M. Preparation and characterization of sodium alginate/poly(*N*-isopropylacrylamide)/clay semi-IPN magnetic hydrogels. *Polym. Bull.* **2012**, *68*, 1153–1169. [CrossRef]
13. Hoare, T.; Pelton, R. Highly pH and temperature responsive microgels functionalized with vinylacetic acid. *Macromolecules* **2004**, *37*, 2544–2550. [CrossRef]
14. Koul, V.; Mohamed, R.; Kuckling, D.; Adler, H.J.; Choudhary, V. Interpenetrating polymer network (IPN) nanogels based on gelatin and poly(acrylic acid) by inverse miniemulsion technique: Synthesis and characterization. *Colloids Surf. B* **2011**, *83*, 204–213. [CrossRef] [PubMed]
15. Berger, J.; Reist, M.; Mayer, J.M.; Felt, O.; Peppas, N.A.; Gurny, R. Structure and interactions in covalently and ionically crosslinked chitosan hydrogels for biomedical applications. *Eur. J. Pharm. Biopharm.* **2004**, *57*, 19–34. [CrossRef]
16. Lipatov, Y.S.; Alekseeva, T.T. Phase-separated interpenetrating polymer networks. *Adv. Polym. Sci.* **2007**, *208*, 1–227.
17. Xia, X.; Hu, Z. Synthesis and light scattering study of microgels with interpenetrating polymer networks. *Langmuir* **2004**, *20*, 2094–2098. [CrossRef] [PubMed]
18. Hu, Z.; Xia, X. Hydrogel nanoparticle dispersions with inverse thermoreversible gelation. *Adv. Mater.* **2004**, *16*, 305–309. [CrossRef]
19. Ma, J.; Fan, B.; Liang, B.; Xu, J. Synthesis and characterization of poly(*N*-isopropylacrylamide)/poly(acrylic acid) semi-IPN nanocomposite microgels. *J. Colloid Interf. Sci.* **2010**, *341*, 88–93. [CrossRef] [PubMed]
20. Xing, Z.; Wang, C.; Yan, J.; Zhang, L.; Li, L.; Zha, L. pH/Temperature dual stimuli-responsive microcapsules with interpenetrating polymer network structure. *Colloid. Polym. Sci.* **2010**, *288*, 1723–1729. [CrossRef]
21. Xing, Z.; Wang, C.; Yan, J.; Zhang, L.; Li, L.; Zha, L. Dual stimuli responsive hollow nanogels with IPN structure for temperature controlling drug loading and pH triggering drug release. *Soft Matter* **2011**, *7*, 7992–7997. [CrossRef]
22. Ahmad, H.; Nurunnabi, M.; Rahman, M.M.; Kumar, K.; Tauer, K.; Minami, H.; Gafur, M.A. Magnetically doped multi stimuli-responsive hydrogel microspheres with IPN structure and application in dye removal. *Colloids Surf. A* **2014**, *459*, 39–47. [CrossRef]
23. Jones, C.D.; Lyon, L.A. Synthesis and characterization of multiresponsive core-shell microgels. *Macromolecules* **2000**, *33*, 8301–8306. [CrossRef]
24. Nilsson, C.; Birnbaum, S.; Nilsson, S. Nanoparticle-based pseudostationary phases in CEC: A breakthrough in protein analysis? *Electrophoresis* **2011**, *32*, 1141–1147. [CrossRef] [PubMed]
25. Kwok, M.-H.; Li, Z.-F.; Ngai, T. Controlling the synthesis and characterization of micrometer-sized PNIPAM microgels with tailored morphologies. *Langmuir* **2013**, *29*, 9581–9591. [CrossRef] [PubMed]

26. Ma, G.-H.; Sone, H.; Omi, S. Preparation of uniform-sized polystyrene–polyacrylamide composite microspheres from a wow emulsion by membrane emulsification. *Macromolecules* **2004**, *37*, 2954–2964. [CrossRef]

27. Wei, Y.; Wang, Y.-X.; Wang, W.; Ho, S.V.; Wei, W.; Ma, G.-H. mPEG–PLA microspheres with narrow size distribution increase the controlled release effect of recombinant human growth hormone. *J. Mater. Chem.* **2011**, *21*, 12691–12699. [CrossRef]

28. Wang, Y.-X.; Qin, J.; Wei, Y.; Li, C.-P.; Ma, G.-H. Preparation strategies of thermo-sensitive P(NIPAM-*co*-AA) microspheres with narrow size distribution. *Powder Technol.* **2013**, *236*, 107–113. [CrossRef]

29. Ma, G.-H. Microencapsulation of protein drugs for drug delivery: Strategy, preparation, and applications. *J. Control Release* **2014**, *193*, 324–340. [CrossRef] [PubMed]

30. Myung, D.; Koh, W.; Ko, J.; Hu, Y.; Carrasco, M.; Noolandi, J.; Ta, C.N.; Frank, C.W. Biomimetic strain hardening in interpenetrating polymer network hydrogels. *Polymer* **2007**, *48*, 5376–5387. [CrossRef]

31. Qi, F.; Wu, J.; Fan, Q.-Z.; He, F.; Tian, G.-F.; Yang, T.Y.; Ma, G.-H.; Su, Z.-G. Preparation of uniform-sized exenatide-loaded PLGA microspheres as long-effective release system with high encapsulation efficiency and bio-stability. *Colloid. Surface B* **2013**, *112*, 492–498. [CrossRef] [PubMed]

32. Fuminori, I.; Kimiko, M. Preparation and properties of monodispersed rifampicin-loaded poly (lactide-*co*-glycolide) microspheres. *Colloids Surfaces B Biointerf.* **2004**, *39*, 17–21.

33. Arash, K.; Rassoul, K. Effect of Alyssum homolocarpum seed gum, Tween 80 and NaCl on droplets characteristics, flow properties and physical stability of ultrasonically prepared corn oil-in-water emulsions. *Food Hydrocolloids* **2011**, *25*, 1149–1157.

34. Ma, G.-H.; Yang, J.; Lv, P.-P.; Wang, L.-Y.; Wei, W.; Tian, R.; Wu, J.; Su, Z.-G. Preparation of uniform microspheres and microcapsules by modified emulsification process. *Macromol. Symp.* **2010**, *288*, 41–48. [CrossRef]

35. Xiao, X.; Zhuo, R.; Xu, J.; Chen, L. Effects of reaction temperature and reaction time on positive thermosensitivity of microspheres with poly(acrylamide)/poly(acrylic acid) IPN shells. *Eur. Polym. J.* **2006**, *42*, 473–478. [CrossRef]

36. Burillo, G.; Briones, M.; Adem, E. IPN's of acrylic acid and N-isopropylacrylamide by gamma and electron beam irradiation. *Nucl. Instrum. Methods Phys. Res. Sect. B* **2007**, *265*, 104–108. [CrossRef]

37. Djonlagi, J.; Petrovi, Z.S. Semi-interpenetrating polymer networks composed of poly(N-isopropyl acrylamide) and polyacrylamide hydrogels. *J. Polym. Sci. Part B* **2004**, *42*, 3987–3999. [CrossRef]

38. Zhao, S.-P.; Ma, D.; Zhang, L.-M. New semi-interpenetrating network hydrogels: Synthesis, characterization and properties. *Macromol. Biosci.* **2006**, *6*, 445–451. [CrossRef] [PubMed]

39. Potorac, S.; Popa, M.; Verestiuc, L.; le Cerf, D. New semi-IPN scaffolds based on HEMA and collagen modified with itaconic anhydride. *Mater. Lett.* **2012**, *67*, 95–98. [CrossRef]

40. Williams, R.A.; Peng, S.J.; Wheeler, D.A.; Morley, N.C.; Taylor, D.; Whalley, M.; Houldsworth, D.W. Controlled production of emulsions using a crossflow membrane. *Chem. Eng. Res. Des.* **1998**, *76*, 902–910. [CrossRef]

41. Mariana, P.-H.; Richard, G.-H. Membrane emulsification for the production of uniform poly-N-isopro pylacrylamide-coated alginate particles using internal gelation. *Chem. Eng. Res. Des.* **2014**, *92*, 1664–1673.

42. John, J.; Klepac, D.; Didović, M.; Sandesh, C.J.; Liu, Y.; Raju, K.V.S.N.; Pius, A.; Valić, S.; Thomas, S. Main chain and segmental dynamics of semi interpenetrating polymer networks based on polyisoprene and poly(methyl methacrylate). *Polymer* **2010**, *51*, 2390–2402. [CrossRef]

43. Liu, T.-Y.; Lin, W.-C.; Huang, L.-Y.; Chen, S.-Y.; Yang, M.-C. Surface characteristics and hemocompatibility of PAN/PVDF blend membranes. *Polym. Adv. Technol.* **2005**, *16*, 413–419. [CrossRef]

44. Thimma Reddy, T.; Takahara, A. Simultaneous and sequential micro-porous semi-interpenetrating polymer network hydrogel films for drug delivery and wound dressing applications. *Polymer* **2009**, *50*, 3537–3546. [CrossRef]

45. Fundueanu, G.; Constantin, M.; Asmarandei, I.; Bucatariu, S.; Harabagiu, V.; Ascenzi, P.; Simionescu, B.C. Poly(N-isopropylacrylamide-*co*-hydroxyethylacrylamide) thermosensitive microspheres: The size of microgels dictates the pulsatile release mechanism. *Eur. J. Pharm. Biopharm.* **2013**, *85*, 614–623. [CrossRef] [PubMed]

46. Khan, A. Preparation and characterization of N-isopropylacrylamide/acrylic acid copolymer core-shell microgel particles. *J. Colloid Interface Sci.* **2007**, *313*, 697–704. [CrossRef] [PubMed]

47. Liu, X.-Y.; Guo, H.; Zha, L.-S. Study of pH/temperature dual stimuli-responsive nanogels with interpenetrating polymer network structure. *Polym. Int.* **2012**, *61*, 1144–1150. [CrossRef]
48. Chen, Y.; Ding, D.; Mao, Z.; He, Y.; Hu, Y.; Wu, W.; Jiang, X. Synthesis of hydroxypropylcellulose-poly(acrylic acid) particles with semi-interpenetrating polymer network structure. *Biomacromolecules* **2008**, *9*, 2609–2614. [CrossRef] [PubMed]
49. Myung, D.; Waters, D.; Wiseman, M.; Duhamel, P.; Noolandi, J.; Ta, C.N.; Frank, C.W. Progress in the development of interpenetrating polymer network hydrogels. *Polym. Adv. Technol.* **2008**, *19*, 647–657. [CrossRef] [PubMed]

polymers

Article

Fabrication of Alkoxyamine-Functionalized Magnetic Core-Shell Microspheres via Reflux Precipitation Polymerization for Glycopeptide Enrichment

Meng Yu [1,†], Yi Di [2,†], Ying Zhang [2], Yuting Zhang [1], Jia Guo [1], Haojie Lu [2,*] and Changchun Wang [1,*]

1   Department of Macromolecular Science, State Key Laboratory of Molecular Engineering of Polymers, Laboratory of Advanced Materials, Fudan University, Shanghai 200433, China; yumeng@fudan.edu.cn (M.Y.); 11110440007@fudan.edu.cn (Yu.Z.); guojia@fudan.edu.cn (J.G.)
2   Institutes of Biomedical Sciences and Department of Chemistry, Fudan University, Shanghai 200032, China; 14111510007@fudan.edu.cn (Y.D.); ying@fudan.edu.cn (Yi.Z.)
*   Correspondence: luhaojie@fudan.edu.cn (H.L.); ccwang@fudan.edu.cn (C.W.); Tel.: +86-021-5423-7618 (H.L.); +86-021-5566-4371 (C.W.)
†   These authors contributed equally.

Academic Editor: Guanghui Ma
Received: 19 January 2016; Accepted: 19 February 2016; Published: 4 March 2016

Abstract: As a facile method to prepare hydrophilic polymeric microspheres, reflux precipitation polymerization has been widely used for preparation of polymer nanogels. In this article, we synthesized a phthalamide-protected *N*-aminooxy methyl acrylamide (NAMAm-*p*) for preparation of alkoxyamine-functionalized polymer composite microspheres via reflux precipitation polymerization. The particle size and functional group density of the composite microspheres could be adjusted by copolymerization with the second monomers, *N*-isopropyl acrylamide, acrylic acid or 2-hydroxyethyl methacrylate. The resultant microspheres have been characterized by TEM, FT-IR, TGA and DLS. The experimental results showed that the alkoxyamine group density of the microspheres could reach as high as 1.49 mmol/g, and these groups showed a great reactivity with ketone/aldehyde compounds. With the aid of magnetic core, the hybrid microspheres could capture and magnetically isolate glycopeptides from the digested mixture of glycopeptides and non-glycopeptides at a 1:100 molar ratio. After that, we applied the composite microspheres to profile the glycol-proteome of a normal human serum sample, 95 unique glycopeptides and 64 glycoproteins were identified with these enrichment substrates in a 5 µL of serum sample.

Keywords: alkoxyamine-functionalized microspheres; reflux precipitation polymerization; magnetic composite microspheres; oxime click; glycoproteins/glycopeptides enrichment

---

1. Introduction

In the past decades, multi-functional polymeric microspheres have attracted great attention because of their broad applications in modern science and technology [1]. The particle size and functional groups of polymeric microspheres both play an important role in practical applications. For example, micron-size microspheres are usually applied as chromatographic separation media in size-exclusion chromatography (SEC) [2,3], and nanoscale microspheres are widely used in biomedical fields [4,5]. In order to fulfil different requirements, more and more hybrid microspheres with inorganic cores and polymeric shells have been prepared with different methods, including emulsion polymerization [6,7], surface initiated living polymerization (e.g., SI-ATRP [8–11], SI-RAFT [12,13]), and so on. The functional microspheres show great potentials in protein enrichment [14–16], drug delivery [17–19], imaging [20,21], and diagnosis [22,23].

Reflux precipitation polymerization (RPP), which is a heterogeneous polymerization system without addition of any surfactants, has been adopted to prepare a variety of microspheres with hydrophilic shell [24,25]. However, depending on the RPP requirement, the available functional monomers are synthetically challenging. Although the post-modification strategy have found success in this regard [26,27], the procedure is tedious and often low yielding. Therefore, exploration of new functional monomers for RPP is highly required.

Imine is an important chemistry structure, which has been studied for many years [28]. The substituent groups on carbon atom affect the stability of the imine structure, and the electrophilic structures can stabilize imine in aqueous solution. Compared with aliphatic Schiff base, the aromatic Schiff base is more stable [29]. Of imine derivatives, hydrazones are very useful in controllable drug release because of their slow hydrolysis property in weak acidic aqueous solution [30]. Oximes are the most stable imine structure, which have been widely used in bioconjugation [31–33]. Recently, it is found that the nucleophilic catalyst could accelerate the reaction between alkoxyamine and ketone/aldehyde groups [34–37]. This finding implies that alkoxyamine-functionalized nanomaterials have great potentials for use in identification of aldehyde-containing biomolecules, for example, glycopeptides and glycoproteins.

Up to date, some papers report the post-modification preparation of alkoxyamine-functionalized polymers [32,35,38], but they do not concern with the problem of functional group density. In order to directly prepare the alkoxyamine polymers or polymer microspheres, *N*-boc-protected alkoxyamine monomers have been synthesized [33,39], but the deprotection condition is not fit to prepare hybrid materials. Sumerlin and co-workers prepared a phthalamide-protected alkoxyamine monomer, and the well-defined polymer was synthesized via RAFT polymerization [40]. However, it is evident that these monomers are not suitable to polymerize in RPP.

Within the context, we designed a new alkoxyamine-based monomer to directly prepare alkoxyamine-functionalized polymer microspheres with magnetite nanoclusters in core by the RPP route. Phthalamide-protected *N*-aminooxy methyl acrylamide (NAMAm-*p*) was synthesized as monomer [41], and it was subjected to the RPP for well controlling the particle size and functional group density. Finally, the tailor-made microspheres were applied in enrichment of glycoproteins and glycopeptides.

2. Materials and Methods

2.1. Materials

Iron(III) chloride hexahydrate (FeCl$_3 \cdot$6H$_2$O), ammonium acetate (NH$_4$Ac), sodium acetate anhydrous (NaAc), ethylene glycol, anhydrous ethanol, trisodium citrate dihydrate, aqueous ammonia solution (NH$_3 \cdot$ H$_2$O, 25%), *N,N'*-dimethylformamide (DMF), aniline, 2,2-azobisisobutyronitrile (AIBN), *N*-isopropyl acrylamide (NIPAm), acrylic acid (AA), 2-hydroxyethyl methacrylate (HEMA) were purchased from Sinopharm Chemical Reagent Co., Ltd. (Shanghai, China). *N,N'*-Methylene-bis-acrylamide (MBA), *N*-methylolacrylamide, diisopropyl azodicarboxylate (DIPA), triphenylphosphine, *N*-hydroxyphthalimide (NOP), hydrazine monohydrate were purchased from Aladdin (Shanghai, China). Methacryloxypropyltriethoxysilane (MPS), bovine serum albumin (BSA), asialofetuin from fetal calf serum (ASF), myoglobin from horse heart (MYO), lysozyme (LYS), sodium periodate (NaIO$_4$), ammonium bicarbonate (NH$_4$HCO$_3$), urea, MALDI matrix (α-cyano-4-hydroxycinnamic acid, CHCA) were all obtained from Sigma (St. Louis, MO, USA). Acetonitrile (ACN, 99.9%, chromatographic grade) and trifluoroacetic acid (TFA) were purchased from Merck (Darmstadt, Germany). The glycerol free peptide-*N*-glycosidase (PNGase F, 500 units/μL) and SDS-PAGE molecular weight standards (6.5–175 kDa) were from New England Biolabs (Ipswich, MA, USA). Sep-Pak C18 columns were from Waters (Shanghai, China). Human serum was provided by Fudan University Shanghai cancer center and stored at −80 °C before analysis. Water used in experiments was ultrapure water prepared using a Milli-Q50SP Reagent Water System (Millipore, Bedford, MA, USA).

2.2. Instrument and Analysis

The polydispersity index (M_w/M_n) of the polymers was measured by gel permeation chromatography (GPC). The GPC anylysis was performed on an Agilent 1100 equipped with a G1310A pump, a G1362A refractive index detector, and a G1315A diode-array detector, poly (methyl methacrylate) (PMMA) standard samples were used as calibration, DMF was used as mobile phase, the measurement condition is at 40 °C with an elution rate of 0.5 mL/min. Transmission electron microscopy (TEM) images were taken on a JEM-2100F transmission electron microscope (JEOL, Tokyo, Japan) at an accelerating voltage of 200 kV. Samples dispersed at an appropriate concentration were cast onto a carbon coated copper grid. Magnetic characterization was carried out on a VSM on a Model 6000 physical property measurement system (Quantum, Blaien, WA, USA) at 300 K. Hydrodynamic diameter (Dh) measurements were conducted by dynamic light scattering (DLS) with a ZEN3600 (Malvern, Malvern, UK) Nano ZS instrument using He–Ne laser at a wavelength of 632.8 nm. Fourier transform infrared spectra (FT-IR) were recorded on a Magna-550 (Nicolet, Waltham, MA, USA) spectrometer. Spectra were scanned over the range of 400–4000 cm^{-1}. All of the dried samples were mixed with KBr and then compressed to form pellets. Thermogravimetric analysis (TGA) measurements were performed on a Pyris 1 TGA instrument (PerkinElmer, Waltham, MA, USA). All measurements were taken under a constant flow of atmosphere of 40 mL/min. The temperature was first increased from room temperature to 100 °C and held until constant weight, and then increased from 100 to 600 °C at a rate of 20 °C/min. NMR data were measured by Varian Mercury AS400 NMR System (Varian, Palo Alto, CA, USA). The sodium dodecyl sulfate–polyacrylamide gel electrophoresis (SDS–PAGE) was performed using 4%–15% precast polyacrylamide gels and Mini-Protean Tetra cell (Tanon, Shanghai, China). Protein concentration was obtained by measuring absorbance at 595 nm using BioTek Power Wave XS2 microplate reader (BioTek, Winooski, VT, USA).

2.3. Synthesis of N-Aminooxy Methylacylimde-p (NAMAm-p)

In a typical procedure, N-methylol acrylimide (10.1 g), N-hydroxyphthalimide (16.3 g), triphenylphosphine (26.2 g) were dissolved by 150 mL ACN in a 250 mL three-necked round-bottomed flask. After purging nitrogen for 0.5 h, diisoproply-azodicarboxylate (20.2 g) was dropped slowly into the solution within 1 h and the solution was stirred overnight at room temperature. At last, 5 mL ethanol was added into the solution and the mixture was concentrated under reduced pressure. The product was washed with 50 mL chloroform for 3 times and dried in the vacuum to obtain the pure NAMAm-p white powder (16.5 g, 65%). ^{1}H NMR (400 Hz, d$_6$-DMSO): δ 9.30 (t, H), δ 7.84 (s, 4H), δ 6.10 (qd, 2H), δ 5.67 (dd, H), δ 5.17 (d, 2H). ^{13}C NMR: δ 165.20 (CONH), δ 164.78 (CONO), δ 135.20 (COCCH), δ132.56 (CHCH$_2$), δ 129.41 (CCHCH), δ 126.51 (CH$_2$CH), δ 123.64 (CHCHCH), δ 63.04 (NHCH$_2$).

2.4. Preparation of PNAMAm-p Polymer Chain and Deprotection for PNAMAm

The PNAMAm polymer was prepared by traditional radical polymerization. Typically, NAMAm-p (500 mg) and AIBN (20 mg, 0.122 mmol) were dissolved by 50 mL ACN in a dried 100 mL two-necked flask, followed by 5 min ultra-sonication to ensure the formation of homogeneous solution. After purging nitrogen for 0.5 h, the solution was stirred with magnetic stirring bar and heated to 90 °C, keeping reflux for 4 h. The product was concentrated under reduced pressure, and the residual powder were dissolved in 20 mL DMF and precipitated with methanol twice. At last, the pure PNAMAm-p white powder (362 mg, 70%) was obtained.

For the deprotection, PNAMAm-p (300 mg) was dissolved in the solution of 20 mL DMF with 2 mL hydrazine hydrat, then the solution was stirred at room temperature for 2 h. The final product of PNAMAm as white powder were collected by precipitation with methanol and dried at 40 °C in vacuum oven (97 mg, 69%).

2.5. Preparation of Monodisperse Crosslinked PNAMAm-p Microspheres and Deprotection Study

The microspheres preparation was carried out via reflux-precipitation polymerization by varying the crosslinker and monomer concentrations and reaction times. A typical polymerization procedure was as follows: NAMAm-*p* (100 mg), MBA (25 mg) and AIBN (5 mg) were dissolved by 20 mL ACN in a dried 50 mL single-necked flask, followed by 5 min ultra-sonication. The flask was then equipped with an allihn condenser and immersed in oil bath, and the reaction temperature was slowly increased from ambient temperature to 90 °C, the reaction solution kept refluxing for one hour without stirring at this temperture. The final products were collected by centrifugation following by repeated washing with ACN and DMF. The crosslinked PNAMAm-*p* microspheres were deprotected with 10% hydrazine hydrate in DMF solution for 2 h at room temperature. The final microspheres were collected by centrifugation and dried at 40 °C in vacuum oven.

2.6. Preparation of MPS Modified MSPs (MSP-m)

The MSPs (magnetic supraparticles) were prepared by a modified solvothermal route [14]. Typically, $FeCl_3 \cdot 6H_2O$ (4.3 g), NaAc (4.8 g), sodium citrate (1.0 g) were dissolved in 70 mL of ethylene glycol. The mixture was stirred vigorously for 1 h at 160 °C to form a homogeneous black solution and then transferred into a Teflon-lined stainless-steel autoclave (100 mL capacity). The autoclave was heated at 200 °C and maintained for 20 h, then it was cooled to room temperature. The black precipitate MSPs were washed twice by ethanol. Then, the MSPs were modified with MPS through a sol–gel method. Typically, all of the products were dispersed in 160 mL ethanol and 40 mL DI water, then 3 mL ammonium hydroxide and 1.2 mL MPS were added into the mixture, and the mixture was stirred overnight at room temperature. The final products were washed twice with ethanol by a magnet, and dried in the vacuum at 40 °C.

2.7. Preparation of MSP@PNAMAm-p Magnetic Hybrid Microspheres and Deprotection for MSP@PNAMAm

The MSP@PNAMAm-*p* magnetic composite microspheres were prepared by a modified RPP route under different conditions by varying the monomer and crosslinker concentrations, in all the formulation, the MSP-m concentration was fixed at 1.25 mg/mL. Typically, MSP-m (25 mg), NAMAm-*p* (100 mg), MBA (25 mg), AIBN (5 mg) were dispersed in 20 mL ACN in a dried single-necked flask by 5 min ultra-sonication to ensure the formation of a stable dispersion. The mixture was heated to 90 °C and kept the reaction for 2 h, the final product was collected and washed with DMF twice. The deprotection procedure was the same as above and the final products were dried at 40 °C in the vacuum for further use.

2.8. Enrichment of Model N-glycoprotein and N-glycopeptides with MSP@PNAMAm Microspheres

In a typical procedure, 5 µg BSA, 5 µg RNase B and 5 µg LYS were dissolved in 40 µL oxidation acetate buffer (pH = 5.6, 100 mM), and then 10 µL 50 mM $NaIO_4$ solution was added. After 1h shaking in the darkness at room temperature, the oxidation process was quenched by 10 µL 50 mM sodium sulfite solution Then, the oxidized products were lyophilized and the coupling buffer (pH = 4.6, containing 10 mM ammonium acetate) were introduced. After the addition of MSP@PNAMAm and 100 mM aniline, the mixture above were kept at 45 °C for 4 h under constant shaking in the dark. Then, the nonspecifically captured peptides were removed by washing twice with 50% DMF and 50 mM NH_4HCO_3 sequentially with the aid of magnet. Next, the microspheres were incubated in the mixture containing 1 µL of PNGase F (500 units per µL) and 50 mM NH_4HCO_3 at 37 °C overnight, the supernatant and eluate were analyzed by SDS–PAGE individually.

For the enrichment of *N*-glycopeptides, the standard protein (ASF and MYO) were dissolved in 50 mM NH_4HCO_3 (pH = 8.0) and denatured by incubating at 100 °C for 10 min. After cooling down to room temperature, trypsin was added to the solution at an enzyme-to-substrate ratio of 1:50 (*w/w*) and the proteolysis was proceeded overnight at 37 °C, followed by lyophilization of the digested sample. The lyophilized glycopeptides and non-glycopeptides at different ratios (1:10, 1:50, 1:100) were used as the enrichment sample, and the procedure of the enrichment was similar to that of model proteins. After digestion of the PNGase F, the supernatant and eluate were analyzed by MALDI-TOF. All experiments of standard glycopeptides were performed in reflector positive mode on AB Sciex 5800 MALDI-TOF/TOF mass spectrometer (AB Sciex, Framingham, MA, USA) with a pulsed Nd/YAG laser at 355 nm. 0.5 μL aliquot of the eluate and 0.5 μL of CHCA (10 mg/mL CHCA in 50% ACN containing 0.1% TFA) matrix were spotted onto a MALDI target plate.

2.9. Enrichment of N-glycopeptides from Human Serum with MSP@PNAMAm

In order to evaluate the enrichment capability of the MSP@PNAMAm magnetic hybrid microspheres, we further used MSP@PNAMAm to enrich glycopeptides in human serum from normal volunteers. The tryptic digest of 5 μL human serum was treated according to the above procedure after reduction and alkylation. Then the eluate was collected and sent for nano-LC-MS/MS analysis. The human serum protein sample was analyzed by a LC-20AD system (Shimadzu, Tokyo, Japan) connected to a LTQ orbitrap mass spectrometer (Thermo Electron, Bremen, Germany) equipped with an online nanoelectrospray ion source (Michrom Bioresources, Auburn, CA, USA). The lyophilized deglycosylated peptides were redissolved in solution containing 5% ACN containing 0.1% FA. Then the sample solution was loaded on a CAPTRAP column (0.5 mm × 2 mm, MICHROM Bioresources, Auburn, CA, USA) in 4 min with a flow rate of 20 μL/min. For a gradient separation, the gradient elution was performed as follows: Acetonitrile from 5% to 45% (95% ACN in 1% FA) over 100 min at a flow rate of 500 nL/min. For each cycle of duty, full mass scan was acquired from 400 to 2000 m/z. The MS/MS spectra were obtained in data-dependent ddMS2 mode. The 12 most intense ions with charge 2, 3 or 4 were selected for the MS/MS run, and the a dynamic exclusion duration was 90 s. Finally, three parallel enrichment operations were performed as technical repeats.

2.10. Data Analysis

For the enrichment of glycopeptides in the human serum, the raw data derived from the ESI MS/MS analysis was searched by MASCOT, against a database (uniprot. Human). The parameters of the search were set as follows: Enzyme of trypsin (partially enzymatic, two missed cleavages were allowed). Fixed modifications of carboxamidomethylation (C, 57.02150), variable modifications of oxidation (M, 15.99492) and deglycosylation (N, 0.98402). 20 ppm error tolerance of precursor mass and 1 Da offragment mass for the Mascot search. Significance threshold was controlled as p value below 5%. Only peptides' sequence containing N-X-S/T(XX-S were considered as *N*-linked glycolpeptides.

3. Results and Discussion

3.1. Polymerization of PNAMAm-p and the Deprotection Study

The monomer NAMAm-*p* was synthesized via Mitsunobu reaction (Scheme 1a) and its molecular structure was confirmed by ^{1}H NMR (Figure S1a, Supplementary Material). Then the PNAMAm was prepared by traditional free radical polymerization of NAMAm-*p* and one-step deprotection (Scheme 1b). Compared with the ^{1}H NMR spectrum of NAMAm-*p*, the PNAMAm-*p* (Figure 1a) gives the peaks at 9.30, 5.17 and 7.84 ppm, which could be ascribed to secondary amine, methylene and phthalimide, respectively, while the peaks of the –CH=CH$_2$ protons (6.10 and 5.67 ppm) was not observed. The result confirmed the obtained PNAMAm-*p* structure, and also, it well agreed with that found in the FT-IR spectra, wherein the stretching peaks of C=C at 1633 cm^{-1} disappeared due to the polymerization of NAMAm-*p* monomer (Figure 1f–I, II).

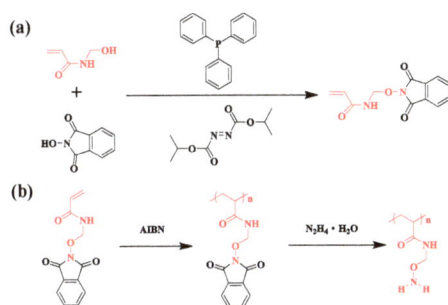

Scheme 1. Synthesis of (**a**) monomer NAMAm-*p* and; (**b**) polymer PNAMAm.

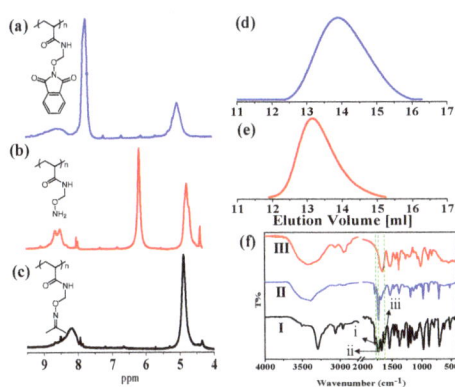

Figure 1. ^{1}H NMR spectra of (**a**) PNAMAm-*p*; (**b**) PNAMAm, and; (**c**) PNAMAm-m; GPC spectra of; (**d**) PNAMAm-*p* (PDI = 1.36), and; (**e**) PANAMAm (PDI = 1.37); (**f**) FT-IR spetra of (I) NAMAm-*p*; (II) PNAMAm-*p*; and (III) PNAMAm. The labelled peaks are (i) 1790 cm^{-1}; (ii) 1735 cm^{-1} and (iii) 1633 cm^{-1}.

Deprotection of the PNAMAm-*p* was accomplished by hydrazine hydrate in DMF at room temperature for 2 h. As shown in Figure 1b, the phthalimide peak at 7.84 ppm disappears, the methylene peak shifts from 5.17 to 4.50 ppm, and a new peak is found at 5.90 ppm, which is attributed to the –O–NH$_2$ protons. The integral area ratio of peaks at 4.50 and 5.90 ppm is about 1:1, which is well consistent with the theory value. The results reveal that the PNAMAm-*p* is completely deprotected and the alkoxyamine groups are formed. Again, FT-IR spetra (Figure 1f) proved the deprotection due to disappearance of the stretching peaks of C=O at 1790 cm^{-1}, C=C at 1735 cm^{-1}. The high efficient deprotection ensures the precise control of functional group density on the polymers. According to GPC measurement (Figure 1d), the M_n of PNAMAm-*p* is 3630 and the M_w is 3940, the PDI is 1.36. A small difference in PDI (Figure 1d,e) was found, implying the molecular weight distribution of polymer chains did not change much in the deprotection process, and only the side groups were changed. However, the GPC results could not provide more information because the polymer polarity was greatly changed.

We estimated the reactivity of the side alkoxyamine groups by the model reaction. Acetone was used to react with PNAMAm to form oxime bonds. As shown in Figure 1c, the peak of –O–NH$_2$ protons at 5.90 ppm disappears, and the peak of methylene protons shifts from 4.50 to 4.90 ppm. This change is owing to the formation of oxime bonds. The product yield, which could be calculated by comparing the integration of the peak at 4.90 ppm, was about 95%. It means that the acetone is efficiently linked to the PNAMAm under the mild conditions. The results demonstrate that the

PNAMAm not only possesses a high reaction activity, but also provides possibility for bioconjugation with aldehyde/ketone-based glycoproteins and glycopeptides in complex physiological environment.

3.2. Preparation of PNAMAm Microspheres and MSP@PNAMAm Core-Shell Microspheres

To satisfy different potential applications, we studied the copolymerization of the monomer NAMAm-*p* with crosslinker and other monomers to fabricate different kinds of alkoxyamine functional microspheres. The detailed design was shown in Scheme 2. First, we investigated the effect of crosslinker species and relative amounts on the microsphere properties. The three often used crosslinkers, divinylbenzene (DVB), ethylene glycol dimethylacrylate (EGDMA) and *N*,*N'*-methylene bisacrylamide (MBA), were utilized for the preparation of PNAMAm-*p* microspheres. The results showed that only MBA could afford the corresponding microspheres, the possible reason is related to the higher reactivity of MBA.

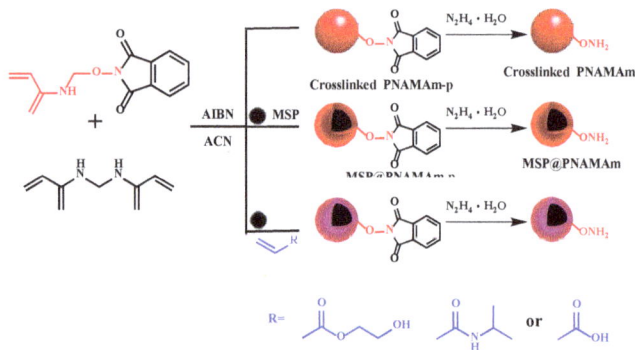

Scheme 2. Preparation of the crosslinked PNAMAm micropsheres, MSP@PNAMAm core–shell microspheres, and the core-shell microspheres with varying functional groups (MSP@PNAMAm-R).

TEM images in Figure 2 display that the particle size is greatly influenced by the feed amount of crosslinker MBA. When 20 wt % MBA was used, the particle size was about 300 nm. With the increase of MBA content to 30 wt % and 50 wt %, the particle sizes accordingly decreased to 200 and 150 nm. Meanwhile, the amount of alkoxyamine groups in the microspheres were reduced (Figure S2, Supplementary Material). When the MBA content was less than 20 wt %, no microspheres were formed.

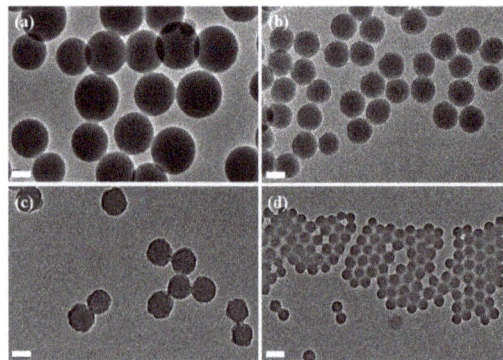

Figure 2. TEM images of PNAMAm with different amount of MBA as crosslinker: (**a**) 20%; (**b**) 30%; (**c**) 40%; (**d**) 50%. The scale bar is 200 nm.

On the basis of the above results, the core–shell magnetic microsphere consisting of a crosslinking PNAMAm-*p* in shell (20% MBA) and a magnetite supraparticle (MSP) in core was successfully prepared (Figure 3). The characteristic peaks at 1790 and 1735 cm^{-1} were found in the FT-IR spectrum for the MSP@PNAMAm-*p*, proving the formation of PNAMAm-*p* component (Figure S3, Supplementary Material). After the deprotection, the MSP@PNAMAm was obtained, and the core-shell structure was preserved, without any change in morphology (Figure 3c).

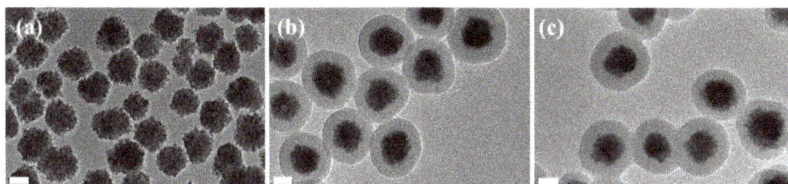

Figure 3. TEM images of (**a**) MSP (magnetic supraparticle); (**b**) MSP@PNAMAm-*p*; (**c**) MSP@PNAMAm. The scale bar is 100 nm.

Meanwhile, TGA results were applied to analyze the compositions, and the density of alkoxyamine group (*d*) could be calculated by the formula below:

$$d = (W_1 - W_2)/(W_2 \times M_{pt})$$ (1)

where, W_1 and W_2 represent the final residual weight percentages of the core-shell composite microspheres after and before the deprotection at 600 °C in air; M_{pt} is 132 g/mol, which is the differ molecular weight caused by deprotection. The microspheres remained 65.4% mass and 78.3% mass at 600 °C before and after deprotection, respectively (Figure 4a). According to the formula above, the alkoxyamine group density was calculated to be 1.49 mmol/g (More details about the calculation are provided in Supplementary Material).

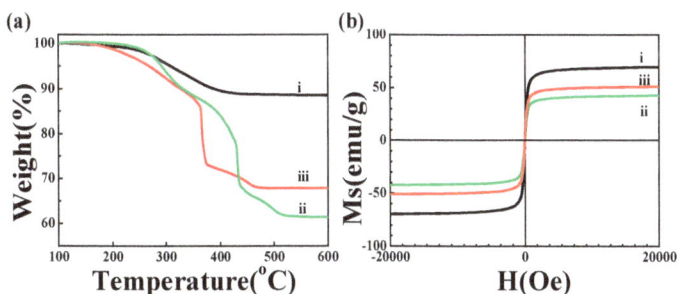

Figure 4. (**a**) TGA and; (**b**) VSM curves of (i) MSP; (ii) MSP@PNAMAm-*p*; (iii) MSP@PNAMAm.

The VSM results also could give the mangetic contents of the resultant microspheres, which coincided with that obtained from the TGA results. Besides, the magnetic hysteresis curves proved that the supraparamagnetic characteristic was unchanged during the deprotection process (Figure 4b).

In our experiment, we found that the solid content played a key role in adjusting the thickness of the polymer shell (Figure 5). When the solid content was 0.25%, only ~5 nm polymer shell could be obtained. When the solid content was 0.375%, the shell thickness increased to 30 nm. With continuous increase of the solid content to 0.675%, the shell thickness could be changed to almost 100 nm. This increment trend was also confirmed by DLS (Figure 5e).

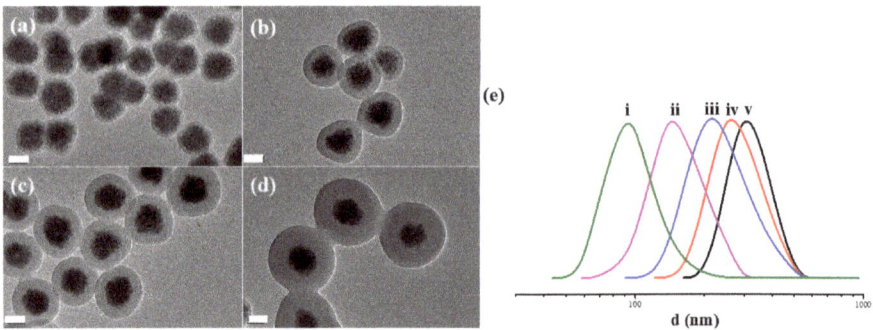

Figure 5. TEM images of MSP@PNAMAm with different thickness of polymeric shell prepared with different solid contents of (**a**) 0.25%; (**b**) 0.375%; (**c**) 0.5%; (**d**) 0.625%, the scale bar is 100 nm; (**e**) DLS results of (i) MSP and MSP@PNAMAm prepared with different solid contents; (ii) 0.25%; (iii) 0.375%; (iv) 0.5%; (v) 0.625%.

In order to quantitatively measure the density of alkoxyamine group in the shell of magnetic composite microspheres, a series of samples with different crosslinking densities were tested by TGA (Table 1). According to the analysis results, the magnetic composite microspheres, which were prepared with different feeding ratios of NAMAm-*p* and MBA and with the same solid content (0.5%), had almost similar shell thickness (Figure S4, Supplementary Material). Through the comparison of the weight changes before and after deprotection, the alkoxyamine group densities were calculated to be 1.49, 1.12, 0.95 and 0.77 mmol/g for the MSP@PNAMAm-1, MSP@PNAMAm-2, MSP@PNAMAm-3, and MSP@PNAMAm-4, respectively.

Table 1. Recipe, TGA data and the calculated alkoxyamine group density of the MSP@PNAMAm samples with different feeding amount of MBA.

Sample	$m_{(MSP)}$ mg	$m_{(NAMAm-p)}$ mg	$m_{(MBA)}$ mg	W_1	W_2	D mmol/g
MSP@PNAMAm-1	25	80	20	78.30%	65.40%	1.49
MSP@PNAMAm-2	25	75	30	72.40%	63.10%	1.12
MSP@PNAMAm-3	25	60	40	76.50%	67.96%	0.95
MSP@PNAMAm-4	25	50	50	73.80%	67.00%	0.77

In addition, the copolymerization properties of NAMAm-*p* were also investigated. NIPAm, AA, and HEMA were used as the second monomers. From the TEM images in Figure 6, the monomer NIPAm has the best performance in copolymerization with NAMAm-*p*, resulting in a uniform and thick polymer shell. The relative ratio of PNAMAm-*p* and PNIPAm could be adjusted by varying the feeding monomer ratios, which was also confirmed by FT-IR (Figure S5, Supplementary Material). Since the monomers of AA and HEMA tend to polymerize on their own, the resultant polymer shells were very thin (Figure 6b,c), but FT-IR spectra (Figures S6 and S7, Supplementary Material) validated that the shells were composed of PNAMAm-*p-co*-PAA and PNAMAm-*p-co*-HEMA copolymers, respectively.

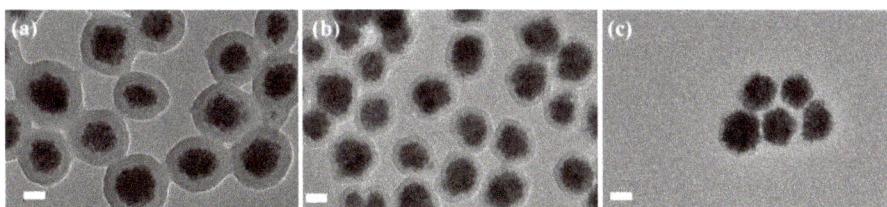

Figure 6. TEM images of (**a**) MSP@PNAMAm-*p-co*-PNIPAm; (**b**) MSP@PNAMAm-*p-co*-PAA; (**c**) MSP@PNAMAm-*p-co*-HEMA. The scale bar is 100 nm.

3.3. Selective Enrichment of Glycoproteins and Glycopeptides

In order to prove the high reactivity between alkoxyamine and aldehyde groups, we evaluated the ability of MSP@PNAMAm to enrich the model glycoproteins. In general, for glycoproteins, the diol structure of saccharide can be oxidized to aldehyde groups, which can bond with the alkoxyamine group. RNB, a typical mono-*N*-glycosylation protein was chosen as the model glycoprotein and the MSP@PNAMAm-1 as the enrichment substrate. The mechanism of enrichment was shown in Scheme 3 and the results were shown in Figure 7. In Lane 1, it was the typical band of protein RNB. After oxidation and conjugation, all of the RNB disappeared in the supernatant (Lane 2). Then with the catalysis of PNGase F, the deglycosylated RNB was eluted in Lane 3. The glycoprotein band with the different molecular weight from Lane 1 represented the lost saccharide structure of RNB by PNGase F. The selectivity of glycoprotein was investigated subsequently by enriching glycoproteins in a mixture with non-glycoproteins (BSA and LYS). The results were shown in Figure 7, Lane 4 to Lane 6. Before enrichment, all the proteins were shown in Lane 4. After oxidation and conjugation, BSA and LYS still remained in the supernatant, but the RNB disappeared in Lane 5. In Lane 6, only deglycosylated RNB was found. There results prove that the MSP@PNAMAm could be applied to enrich glycoproteins selectively, and this property provides a potential ability to enrich glycoproteins and glycopeptides selectively in a complex system.

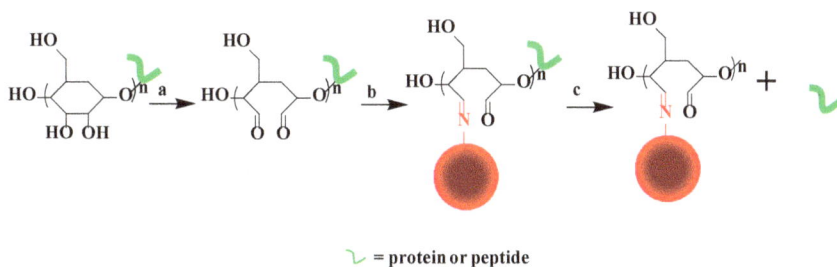

ᒐ = protein or peptide

(a) oxidation by NaIO₄; (b) conjugation by MSP@PNAMAm; (c) deglycosylation by PNGase F.

Scheme 3. Mechanism of glycoprotein and glycopeptide enrichment with magnetic core–shell microspheres (MSP@PNAMAm-1).

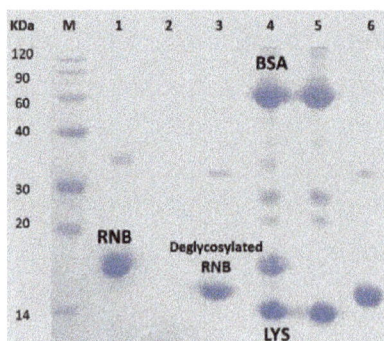

Figure 7. SDS–PAGE analysis of the model glycoprotein proteins before and after treatment with MSP@PNAMAm-1 core-shell microspheres. M stands for protein marker; Lane 1 represents the RNase B; Lane 2 represents the supernatant after enrichment with MSP@PNAMAm-1; Lane 3 represents the released deglycoslated RNB after enrichment; Lane 4 represents the protein mixture of BSA, RNB(RNase B)and LYS(The amount of BSA:RNB:LYS = 1:1:1); Lane 5 represents the supernatant of the protein mixture after enrichment; Lane 6 represents the released deglycosylated RNB after enrichment.

Meanwhile, the selective enrichment of glycopeptides was also investigated from the digests of the mixture containing ASF and MYO (ASF:MYO = 1:10). Considering the effect of functional group density, MSP@PNAMAm-1 and MSP@PNAMAm-4, which had 1.49 and 0.77 mmol/g of alkoxyamine groups, respectively, were applied in glycopeptides enrichment, and the results were shown in Figure 8. Before enrichment, the dominant peaks in the spectrum were due to non-glycopeptides. After enrichment with the two samples, all the dominant peaks (m/z values of 1627.6, 1755.7, 1950.8 and 3017.4) were attributed to deglycopeptides from the digest of ASF. What is more, it was interesting that the glycopeptide distribution in the mass spectrum were affected by the enrichment materials. For the MSP@PNAMAm-4, the strongest peak was at 3017.4 (m/z), while the intensity of this peak was weaker than those at 1627.6 (m/z) and 1755.7 (m/z) with MSP@PNAMAm-1. In addition, it was observed when non-glycopeptide was mixed with glycopeptides in a high molar ratio (100:1) (Figure S8, Supplementary Material). The early reports found the same phenomenon [29,42]. In consideration of the complex conjugation-elution process, it is difficult to explain this phenomenon now, but it is sure that different surface structures have varying interaction with glycopeptides, and the difference might affect the enrichment results.

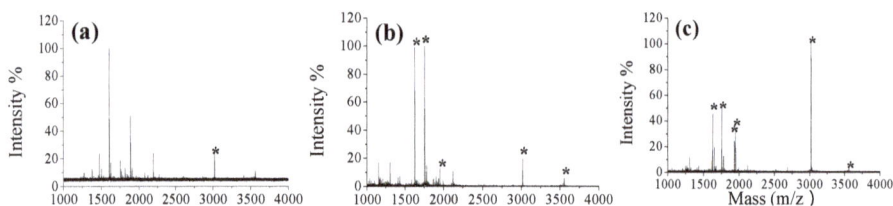

Figure 8. MALDI-TOF mass spectra of the tryptic digest mixture of ASF (asialofetuin) and MYO (myoglobin), the mole ratio of ASF:MYO = 1:10. (**a**) direct analysis; (**b**) analysis after enrichment by MSP@NAMAm-1 and deglycosylation by PNGase F; (**c**) analysis after enrichment by MSP@NAMAm-4 and deglycosylation by PNGase F. The symbols * denote the deglycosylated peptides.

The enrichment ability of MSP@PNAMAm-1 and MSP@PNAMAm-4 were further investigated by profiling the *N*-glycoproteome of a normal human serum sample. We found that the result of MSP@PNAMAm-1 was not good. In contrast, MSP@PNAMAm-4 with the low functional group density showed much better enrichment ability, 95 unique glycopeptides and 64 glycoproteins were identified in a 5 μL serum sample (Table S1, Supplementary Material). It implies that a suitable functional group density is benificial for glycoproteins and glycopeptides enrichment in complex physiological environment. In addition, it proves that the MSP@PNAMAm microspheres exhibit a great potential in real biomedical application.

4. Conclusions

In this paper, we have prepared the alkoxyamine-functionalized magnetic core-shell microspheres via reflux precipitation polymerization. Alkoxyamine acrylamide monomer with a protective group phthalamide was synthesized, and underwent the reflux precipitation polymerization for alkoxyamine-functionalized polymer microspheres. By varying a series of reaction conditions including monomer concentration, comonomer species and crosslinker content, the core-shell magnetic composite microspheres were constructed to achieve the tunable shell thickenss, controllable alkoxyamine group density, and versatile copolymer composition. After the deprotection, the MSP@PNAMAm displayed high activity to couple with carbonyl compounds under mild conditions. They could serve as enrichment substrate to efficiently identify glycoprotein/glycopeptides. Moreover, we found success in profiling the glycoproteome in human serum samples with the structure-optimized composite microspheres.

Supplementary Materials: Supplementary materials can be found at www.mdpi.com/2073-4360/8/3/xx/s1.

Acknowledgments: This work was supported by National Science and Technology Key Project of China (Grants 2012AA020204), NSF (Grants 21474017 and 21335002), Shanghai Projects (Eastern Scholar, and B109)

Author Contributions: Meng Yu and Yi Di designed and performed the experiments and darft the paper; Ying Zhang, Jia Guo and Yuting Zhang discussed the results. Changchun Wang and Haojie Lu conceived and revised the paper.

Conflicts of Interest: The authors declare no conflict of interest.

References

1. Li, G.L.; Möhwald, H.; Shchukin, D.G. Precipitation polymerization for fabrication of complex core–shell hybrid particles and hollow structures. *Chem. Soc. Rev.* **2013**, *42*, 3628–3646. [CrossRef] [PubMed]

2. Barahona, F.; Turiel, E.; Cormack, P.A.G.; Martín-Esteban, A. Chromatographic performance of molecularly imprinted polymers: Core–shell microspheres by precipitation polymerization and grafted mip films via iniferter-modified silica beads. *J. Polym. Sci. Polym. Chem.* **2010**, *48*, 1058–1066. [CrossRef]

3. Qin, W.W.; Silvestre, M.E.; Kirschhöfer, F.; Brenner-Weiss, G.; Franzreb, M. Insights into chromatographic separation using core–shell metal–organic frameworks: Size exclusion and polarity effects. *J. Chromatogr. A* **2015**, *1411*, 77–83. [CrossRef] [PubMed]

4. Mhlanga, N.; Sinha Ray, S.; Lemmer, Y.; Wesley-Smith, J. Polylactide-based magnetic spheres as efficient carriers for anticancer drug delivery. *ACS Appl. Mater. Interfaces* **2015**, *7*, 22692–22701. [CrossRef] [PubMed]

5. Lee, W.L.; Guo, W.M.; Ho, V.H.B.; Saha, A.; Chong, H.C.; Tan, N.S.; Widjaja, E.; Tan, E.Y.; Loo, S.C.J. Inhibition of 3-D tumor spheroids by timed-released hydrophilic and hydrophobic drugs from multilayered polymeric microparticles. *Small* **2014**, *10*, 3986–3996. [CrossRef] [PubMed]

6. Pan, M.R.; Sun, Y.F.; Zheng, J.; Yang, W.L. Boronic acid-functionalized core–shell–shell magnetic composite microspheres for the selective enrichment of glycoprotein. *ACS Appl. Mater. Interfaces* **2013**, *5*, 8351–8358. [CrossRef] [PubMed]

7. Ge, J.P.; Hu, Y.X.; Zhang, T.R.; Yin, Y.D. Superparamagnetic composite colloids with anisotropic structures. *J. Am. Chem. Soc.* **2007**, *129*, 8974–8975. [CrossRef] [PubMed]

8. Zhao, L.; Qin, H.; Hu, Z.; Zhang, Y.; Wu, R.A.; Zou, H. A poly(ethylene glycol)-brush decorated magnetic polymer for highly specific enrichment of phosphopeptides. *Chem. Sci.* **2012**, *3*, 2828–2838. [CrossRef]

9. Qin, W.; Song, Z.; Fan, C.; Zhang, W.; Cai, Y.; Zhang, Y.; Qian, X. Trypsin immobilization on hairy polymer chains hybrid magnetic nanoparticles for ultra fast, highly efficient proteome digestion, facile ^{18}O labeling and absolute protein quantification. *Anal. Chem.* **2012**, *84*, 3138–3144. [CrossRef] [PubMed]

10. Grignard, B.; Jérôme, C.; Calberg, C.; Jérôme, R.; Wang, W.; Howdle, S.M.; Detrembleur, C. Copper bromide complexed by fluorinated macroligands: Towards microspheres by ATRP of vinyl monomers in sc CO_2. *Chem. Commun.* **2008**, *44*, 314–316. [CrossRef]

11. Gao, Y.; Zhou, D.; Zhao, T.; Wei, X.; Mcmahon, S.; Ahern, J.O.; Wang, W.; Greiser, U.; Rodriguez, B.J.; Wang, W. Intramolecular cyclization dominating homopolymerization of multivinyl monomers toward single-chain cyclized/knotted polymeric nanoparticles. *Macromolecules* **2015**, *48*, 6882–6889. [CrossRef]

12. Gonzato, C.; Courty, M.; Pasetto, P.; Haupt, K. Magnetic molecularly imprinted polymer nanocomposites via surface-initiated RAFT polymerization. *Adv. Funct. Mater.* **2011**, *21*, 3947–3953. [CrossRef]

13. Li, X.; Bao, M.M.; Weng, Y.Y.; Yang, K.; Zhang, W.D.; Chen, G.J. Glycopolymer-coated iron oxide nanoparticles: Shape-controlled synthesis and cellular uptake. *J. Mater. Chem. B* **2014**, *2*, 5569–5575. [CrossRef]

14. Zhang, Y.T.; Ma, W.F.; Li, D.; Yu, M.; Guo, J.; Wang, C.C. Benzoboroxole-functionalized magnetic core/shell microspheres for highly specifi c enrichment of glycoproteins under physiological conditions. *Small* **2014**, *10*, 1379–1386. [CrossRef] [PubMed]

15. Zheng, J.; Ma, C.J.; Sun, Y.F.; Pan, M.R.; Li, L.; Hu, X.J.; Yang, W.L. Maltodextrin-modified magnetic microspheres for selective enrichment of maltose binding proteins. *ACS Appl. Mater. Interfaces* **2014**, *6*, 3568–3574. [CrossRef] [PubMed]

16. Zheng, J.; Li, Y.P.; Sun, Y.F.; Yang, Y.K.; Ding, Y.; Lin, Y.; Yang, W.L. A generic magnetic microsphere platform with "clickable" ligands for purification and immobilization of targeted proteins. *ACS Appl. Mater. Interfaces* **2015**, *7*, 7241–7250. [CrossRef] [PubMed]

17. Huang, J.; Shu, Q.; Wang, L.Y.; Wu, H.; Wang, A.Y.; Mao, H. Layer-by-layer assembled milk protein coated magnetic nanoparticle enabled oral drug delivery with high stability in stomach and enzyme-responsive release in small intestine. *Biomaterials* **2015**, *39*, 105–113. [CrossRef] [PubMed]

18. Liu, F.; Wang, J.N.; Cao, Q.Y.; Deng, H.D.; Shao, G.; Deng, D.Y.B.; Zhou, W.Y. One-step synthesis of magnetic hollow mesoporous silica (MHMS) nanospheres for drug delivery nanosystems via electrostatic self-assembly templated approach. *Chem. Commun.* **2015**, *51*, 2357–2360. [CrossRef] [PubMed]

19. Wang, W.; Liang, H.; Racha, C.A.G.; Hamilton, L.; Fraylich, M.; Shakesheff, K.M.; Saunders, B.; Alexander, C. Biodegradable thermoresponsive microparticle dispersions for injectable cell delivery prepared using a single–step process. *Adv. Mater.* **2009**, *21*, 1809–1813. [CrossRef]

20. Huang, L.; Ao, L.J.; Wang, W.; Hu, D.H.; Sheng, Z.H.; Su, W. Multifunctional magnetic silica nanotubes for mr imaging and targeted drug delivery. *Chem. Commun.* **2015**, *51*, 3923–3926. [CrossRef] [PubMed]

21. Nakamura, T.; Sugihara, F.; Matsushita, H.; Yoshioka, Y.; Mizukami, S.; Kikuchi, K. Mesoporous silica nanoparticles for ^{19}F magnetic resonance imaging, fluorescence imaging, and drug delivery. *Chem. Sci.* **2015**, *6*, 1986–1990. [CrossRef]

22. Li, D.; Zhang, Y.T.; Li, R.M.; Guo, J.; Wang, C.C.; Tang, C.B. Selective capture and quick detection of targeting cells with sers-coding microsphere suspension chip. *Small* **2015**, *11*, 2200–2208. [CrossRef] [PubMed]

23. Kim, J.A.; Kim, M.; Kang, S.M.; Lim, K.T.; Kim, T.S.; Kang, J.Y. Magnetic bead droplet immunoassay of oligomer amyloid beta for the diagnosis of alzheimer's disease using micro-pillars to enhance the stability of the oil-water interface. *Biosens. Bioelectron.* **2015**, *67*, 724–732. [CrossRef] [PubMed]

24. Wang, F.; Zhang, Y.; Yang, P.; Jin, S.; Yu, M.; Guo, J.; Wang, C. Fabrication of polymeric microgels using reflux-precipitation polymerization and its application for phosphoprotein enrichment. *J. Mater. Chem. B* **2014**, *2*, 2575–2582. [CrossRef]

25. Chen, Y.; Xiong, Z.; Zhang, L.; Zhao, J.; Zhang, Q.; Peng, L.; Zhang, W.; Ye, M.; Zou, H. Facile synthesis of zwitterionic polymer-coated core–shell magnetic nanoparticles for highly specific capture of N-linked glycopeptides. *Nanoscale* **2015**, *7*, 3100–3108. [CrossRef] [PubMed]

26. Chen, H.; Deng, C.; Yan, L.; Ying, D.; Yang, P.; Zhang, X. A facile synthesis approach to c 8 -functionalized magnetic carbonaceous polysaccharide microspheres for the highly efficient and rapid enrichment of peptides and direct maldi-tof-ms analysis. *Adv. Mater.* **2009**, *21*, 2200–2205. [CrossRef]

27. Zhang, Y.T.; Yang, Y.K.; Ma, W.F.; Guo, J.; Lin, Y.; Wang, C.C. Uniform magnetic core/shell microspheres functionalized with Ni^{2+}-iminodiacetic acid for one step purification and immobilization of his-tagged enzymes. *ACS Appl. Mater. Interfaces* **2013**, *5*, 2626–2633. [CrossRef] [PubMed]

28. Layer, R.W. The chemistry of imines. *Chem. Rev.* **1962**, *63*, 489–510. [CrossRef]

29. Zhang, Y.; Yu, M.; Zhang, C.; Ma, W.F.; Zhang, Y.T.; Wang, C.C.; Lu, H.J. Highly selective and ultra fast solid-phase extraction of *N*-glycoproteome by oxime click chemistry using aminooxy-functionalized magnetic nanoparticles. *Anal. Chem.* **2014**, *86*, 7920–7924. [CrossRef] [PubMed]

30. Li, D.; Tang, J.; Wei, C.; Guo, J.; Wang, S.L.; Chaudhary, D.; Wang, C.C. Doxorubicin-conjugated mesoporous magnetic colloidal nanocrystal clusters stabilized by polysaccharide as a smart anticancer drug vehicle. *Small* **2012**, *8*, 2690–2697. [CrossRef] [PubMed]

31. Niu, J.; Hili, R.; Liu, D.R. Enzyme-free translation of DNA into sequence-defined synthetic polymers structurally unrelated to nucleic acids. *Nat. Chem.* **2013**, *5*, 282–292. [CrossRef] [PubMed]

32. Grover, G.N.; Lam, J.; Nguyen, T.H.; Segura, T.; Maynard, H.D. Biocompatible hydrogels by oxime click chemistry. *Biomacromolecules* **2012**, *13*, 3013–3017. [CrossRef] [PubMed]

33. Mackenzie, K.J.; Francis, M.B. Recyclable thermoresponsive polymer–cellulase bioconjugates for biomass depolymerization. *J. Am. Chem. Soc.* **2013**, *135*, 293–300. [CrossRef] [PubMed]

34. Wendeler, M.; Grinberg, L.; Wang, X.Y.; Dawson, P.E.; Baca, M. Enhanced catalysis of oxime-based bioconjugations by substituted anilines. *Bioconjugate Chem.* **2014**, *25*, 93–101. [CrossRef] [PubMed]

35. Thygesen, M.B.; Munch, H.; Sauer, J.; Clo, E.; Jorgensen, M.R.; Hindsgaul, O.; Jensen, K.J. Nucleophilic catalysis of carbohydrate oxime formation by anilines. *J. Org. Chem.* **2010**, *75*, 1752–1755. [CrossRef] [PubMed]

36. Loskot, S.A.; Zhang, J.J.; Langenhan, J.M. Nucleophilic catalysis of meon-neoglycoside formation by aniline derivatives. *J. Org. Chem.* **2013**, *78*, 12189–12193. [CrossRef] [PubMed]

37. Dirksen, A.; Hackeng, T.M.; Dawson, P.E. Nucleophilic catalysis of oxime ligation. *Angew. Chem.* **2006**, *45*, 7581–7584. [CrossRef] [PubMed]

38. Thygesen, M.B.; Sorensen, K.K.; Clo, E.; Jensen, K.J. Direct chemoselective synthesis of glyconanoparticles from unprotected reducing glycans and glycopeptide aldehydes. *Chem. Commun.* **2009**, 6367–6369. [CrossRef] [PubMed]

39. Shimaoka, H.; Kuramoto, H.; Furukawa, J.-I.; Miura, Y.; Kurogochi, M.; Kita, Y.; Hinou, H.; Shinohara, Y.; Nishimura, S.-I. One-pot solid-phase glycoblotting and probing by transoximization for high-throughput glycomics and glycoproteomics. *Chem. A Eur. J.* **2007**, *13*, 1664–1673. [CrossRef] [PubMed]

40. Hill, M.R.; Mukherjee, S.; Costanzo, P.J.; Sumerlin, B.S. Modular oxime functionalization of well-defined alkoxyamine-containing polymers. *Polym. Chem.* **2012**, *3*, 1758–1762. [CrossRef]

41. Kumara Swamy, K.C.; Bhuvan Kumar, N.N.; Balaraman, E.; Pavan Kumar, K.V.P. Mitsunobu and related reactions advances and applications. *Chem. Rev.* **2009**, *109*, 2551–2651. [CrossRef] [PubMed]

42. Liu, L.; Yu, M.; Zhang, Y.; Wang, C.; Lu, H. Hydrazide functionalized core–shell magnetic nanocomposites for highly specific enrichment of *N*-glycopeptides. *ACS Appl. Mater. Interfaces* **2014**, *6*, 7823–7832. [CrossRef] [PubMed]

polymers

MDPI

Article

Polydopamine Particle as a Particulate Emulsifier

Nobuaki Nishizawa [1], Ayaka Kawamura [2], Michinari Kohri [2], Yoshinobu Nakamura [1] and Syuji Fujii [1,*]

[1] Department of Applied Chemistry, Osaka Institute of Technology, 5-16-1 Omiya, Asahi-ku, Osaka 535-8585, Japan; m1m15509@st.oit.ac.jp (N.N.); yoshinobu.nakamura@oit.ac.jp (Y.N.)

[2] Division of Applied Chemistry and Biotechnology, Graduate School of Engineering, Chiba University, 1-33 Yayoi-cho, Inage-ku, Chiba 263-8522, Japan; a.kawamura@chiba-u.jp (A.K.); kohri@faculty.chiba-u.jp (M.K.)

* Correspondence: syuji.fujii@oit.ac.jp; Tel.: +81-6-6954-4274

Academic Editor: To Ngai
Received: 23 January 2016; Accepted: 18 February 2016; Published: 26 February 2016

Abstract: "Pickering-type" emulsions were prepared using polydopamine (PDA) particles as a particulate emulsifier and n-dodecane, methyl myristate, toluene or dichloromethane as an oil phase. All the emulsions prepared were oil-in-water type and an increase of PDA particle concentration decreased oil droplet diameter. The PDA particles adsorbed to oil–water interface can be crosslinked using poly(ethylene imine) as a crosslinker, and the PDA particle-based colloidosomes were successfully fabricated. Scanning electron microscopy studies of the colloidosomes after removal of inner oil phase revealed a capsule morphology, which is strong evidence for the attachment of PDA particles at the oil–water interface thereby stabilizing the emulsion. The colloidosomes after removal of inner oil phase could retain their capsule morphology, even after sonication. On the other hand, the residues obtained after oil phase removal from the PDA particle-stabilized emulsion prepared in the absence of any crosslinker were broken into small fragments of PDA particle flocs after sonication.

Keywords: Pickering emulsion; polydopamine; oil–water interface; crosslinking; colloidosome

1. Introduction

Emulsions stabilized with solid particles (so-called "Pickering emulsions") have received great interest in the colloid and interface research area [1–5]. In essence, particulate emulsifiers offer more robust, reproducible formulations and lower toxicity profiles compared to conventional molecular-level emulsifiers. It has been well-known that various types of solid particles can work as emulsifier: inorganic particles such as silica [6,7], metals [8–10], semiconductors [11,12], clays [13], or ceramics [14,15]; organic particles such as bionano-particles including viruses [16] and proteins [17–19]; latex particles [20–27]; microgel particles [28–31]; and micelles [32] have been used as an effective particulate emulsifier.

Polydopamine (PDA), a mimetic of mussel adhesive proteins, has attracted much attention as a coating material without surface pre-treatments [33]. Dopamine monomer can be self-polymerized under basic condition on a variety of materials, such as metals, inorganic materials, and polymer materials. Extensive studies have been carried out to create PDA-coated materials with controllable film thickness and stability [34–37]. Some of the present authors have reported the preparation of PDA layers containing atom transfer radical polymerization initiating groups [38,39], polyethylene glycol moieties [40], dyes [41], and carboxylic acid-bearing compounds [42] to produce functional polymeric materials. Another advantage of PDA coating lies in their chemical structures that contain numerous functional groups, such as catechol and amine groups. Because of this advantage, PDA were easily modified by post-functionalization [43] or crosslinking [44]. Although most studies in which a PDA layer has been used have involved modifying the materials' surface, there are a limited number of reports on the preparation of PDA in a form of particles, which were used for metal adsorbent materials [45], anti-cancer drug delivery [46], biomedical applications [47], and carbon source [48]. Unfortunately, the PDA particles synthesized previously are polydisperse in size, and synthesis of PDA particles with high monodispersity are desired. Under these situations, we have succeeded in fabrication of monodisperse PDA particles in water–methanol solution, and their use as bright structural color materials [49].

Herein, we describe the evaluation of PDA particle as a particulate emulsifier for the first time and utilize a liquid–liquid interface as a tool to assemble PDA particles. Thanks to catechol group on the PDA surface, PDA particles assembled at the droplet interface could be subsequently crosslinked using poly(ethylene imine) in order to stabilize these superstructures and to fabricate colloidosomes.

2. Materials and Methods

2.1. Materials

Reagents used to prepare PDA particles were dopamine hydrochloride (DA–HCl, Kanto Chemical, Tokyo, Japan), tris(hydroxymethyl)aminomethane (Tris, Kanto Chemical), and methanol (Kanto Chemical). Oils used to prepare emulsions were *n*-dodecane (≥99%, Sigma-Aldrich, Tokyo, Japan), toluene (99%, Sigma-Aldrich), dichloromethane (DCM, ≥99.0%, Sigma-Aldrich), methyl myristate (95.0%, Wako Pure Chemical, Osaka, Japan), octafluorotoluene (97%, Wako Pure Chemical) and perfluorononane (99%, Wako Pure Chemical). Poly(ethylene imine) (PEI, Average Molecular Weight, approximately 600, Wako Pure Chemical) was used as a crosslinking agent for PDA particles. Poly(vinyl alcohol) (PVA) were purchased from Sigma-Aldrich. Deionized water (<0.06 $\mu S \cdot cm^{-1}$, Advantec MFS RFD240NA: GA25A-0715) was used for preparation of PDA particles and emulsions. All other chemicals and solvents were of reagent grade and used as received.

2.2. PDA Particles Synthesis

DA–HCl (1.7 mg/mL, 10.8 mmol), Tris (14.4 g, 120 mmol), and water/methanol ($w/w = 4/1$) solution (1.2 L) were placed in a flask, and the mixture was stirred at 30 °C for 20 h. The PDA particles were separated and purified repeatedly by centrifugation (10,000 rpm (15,600× g) for 30 min) and redispersed in deionized water.

2.3. PDA Particles Characterization

2.3.1. Scanning Electron Microscopy (SEM) Study

Scanning electron microscopy (SEM; JSM-6510A; JEOL, 20 kV, JEOL, Tokyo, Japan) studies were conducted on Pt sputter-coated (JFC-1600 Auto Fine Coater; JEOL) dried samples.

2.3.2. Dynamic Light Scattering (DLS) and Zeta Potential Studies

The hydrodynamic diameter (D_h; in water) and the zeta potential (in 0.01 M NaCl aqueous solution) of the PDA particles were measured by dynamic light scattering (DLS; ELSZ-1000ZS; Otsuka Electronics Co. Ltd., Osaka, Japan).

2.3.3. Density

The solid-state density of the dried PDA particles was determined by helium pycnometry using a Micromeritics AccuPyc II 1340 instrument (Micromeritics, Norcross, GA, USA).

2.3.4. Infrared (IR) study

IR spectra were measured by IR spectrophotometer (FTIR-420; JASCO, Tokyo, Japan).

2.4. Emulsion Preparation

Each volume (3.0 mL) of aqueous dispersion of the PDA particles with a solid concentration of 1.00 wt % and oil (*n*-dodecane, methyl myristate, toluene or dichloromethane) were placed in a glass vessel (inner volume, 13 mL). The two phases were homogenized for 2 min at 25 °C using a homogenizer (IKA ULTRA-TURRAX® T 25 digital, Staufen, Germany) equipped with a dispersing element (S25N-8G: stainless steel, 18 mm stator diameter, 12.7 mm rotor diameter, 108 mm shaft length) operating at 20,000 rpm. Emulsion stabilities after standing for 24 h at 25 °C were assessed by gravimetric or visual inspection.

2.5. Crosslinking of Particle-Stabilized Emulsion

PEI aqueous solution (5.0 mL, 0.05–10 wt %) was added to the PDA particle-stabilized DCM-in-water emulsion (1.0 mL, prepared at a PDA particle concentration of 1.00 wt %) and magnetically stirred to ensure homogeneous mixing. The emulsion was then allowed to stand stirred at 25 °C for 1 h to allow colloidosome formation to occur.

2.6. Sonication Challenge

An aliquot of Pickering emulsion or colloidosome sample prepared using DCM (0.40 mL) was purified by five times replacement of supernatant with deionized water, and then was sonicated using an ultrasonic washing machine (Bransonic 221, Yamato Scientific Co., Tokyo, Japan) for 1 h. During the purification, DCM oil phase was removed from the droplets because of dissolution to repeatedly replaced water media. The sample was viewed by both optical microscopy and SEM.

2.7. Emulsion Characterization

2.7.1. Drop Test

Emulsion type was confirmed using "drop test". One drop of the emulsion was added to both water and oil and its ease of dispersion was assessed by visual inspection. Relatively rapid dispersion indicated that the continuous phase of the emulsion was the same as the diluent.

2.7.2. Optical Microscopy (OM) Study

A drop of the diluted emulsion was placed on a microscope slide and viewed using an optical microscope (Motic BA200, Shimadzu, Kyoto, Japan) fitted with a digital system (Moticam 2000, Shimadzu).

2.7.3. Laser Diffraction Study

A Malvern Mastersizer2000 instrument equipped with a small volume Hydro 2000 SM sample dispersion unit (*ca.* 150 mL including flow cell and tubing), a HeNe laser operating at 633 nm and solid-state blue laser source operating at 466 nm were used to size the emulsion droplets. The stirring rate was adjusted to 2000 rpm in order to avoid creaming of the emulsion. It was confirmed that droplet size did not change under these measurement conditions, which indicated no coalescence of the emulsion droplets. The raw data were analyzed using Malvern software. The mean droplet diameter was taken to be the volume average diameter ($D_{4/3}$), which is mathematically expressed as $D_{4/3} = \Sigma D_i^4 N_i / \Sigma D_i^3 N_i$ (D_i, the diameters of individual droplets; N_i, the number of emulsion droplets corresponding to the diameters). The $D_{4/3}$ values were shown plus–minus standard deviations, which were also determined using the Malvern software (Malvern Instruments, Malvern, UK). Droplet size can be measured from 0.02 to 2000 μm. Light diffraction method is an authorized technique to measure mean droplet diameters and their distributions. There is a high reproducibility in light diffraction measurements in our study.

2.7.4. Scanning Electron Microscopy Study

Scanning electron microscopy (SEM; Keyence VE-8800, 12 kV, Keyence, Osaka, Japan) studies were conducted on Au sputter-coated (Elionix SC-701 Quick Coater, Sanyu Electron, Tokyo, Japan) dried samples.

3. Results and Discussion

3.1. PDA Particles Characterization

First, the PDA particles obtained were characterized by FT-IR spectroscopy (Figure 1a). The IR spectrum of the PDA particles shows a broad peak at 3200–3500 cm^{-1} due to the hydroxyl structures such as a catechol group. The characteristic peaks of indole and indoline structures at approximately 1600 cm^{-1} and approximately 1500 cm^{-1}, respectively, were also found in PDA particles. Figure 1b shows a digital photograph of the PDA particle dispersion (0.5 wt % in water), which confirms the formation of black-colored aqueous dispersion. The hydrodynamic diameter, measured by DLS, was approximately 220 ± 39 nm (Figure 1c). The zeta potential of the PDA particles was measured to be approximately −42 mV (in 0.01 M NaCl aqueous solution), and the PDA particles with negative surface charges were well dispersed in water with no flocs were observed over a one-month period. The PDA particles size used in this study was measured to be 220 nm from the SEM image (Figure 1d), which accords well with that reported previously [49]. As indicated in the SEM image, PDA particles obtained are near monodisperse, which is consistent with the result obtained using DLS, and have smooth spherical surface. The PDA particles were used as a particulate emulsifier in their colloidally stable state.

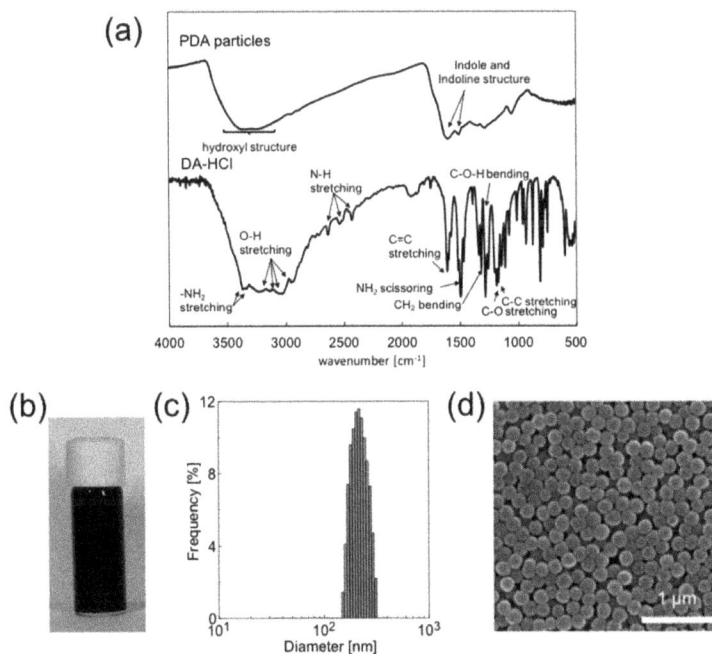

Figure 1. (**a**) FT-IR spectra of PDA particles and DA–HCl; (**b**) digital photograph of an aqueous dispersion of PDA particles (0.50 wt %); (**c**) size distribution of PDA particles measured by DLS; and (**d**) SEM image of PDA particles.

3.2. Emulsions Stabilized with PDA Particles

3.2.1. Different Oils Emulsion Data

In order to check the ability of the PDA particles as a Pickering-type emulsifier, four oils, namely *n*-dodecane, methyl myristate, toluene and DCM, were used as an oil phase to prepare emulsions at a PDA concentration of 1.00 wt %. *n*-Dodecane and methyl myristate are non-volatile at room temperature, and toluene and dichloromethane are volatile and used in order to characterize droplet after evaporation of oil phase in detail. In all cases, highly stable oil-in-water emulsions were achieved after homogenization and OM studies revealed that oil droplets stably dispersed in aqueous continuous phase without coalescence. All emulsions prepared in this study (beside methyl myristate) survived at least one month and nearly 100% emulsions remained in closed system where the evaporation of oil and water are not allowed: in the methyl myristate system, 39% demulsification occurred after two months.

Homogenization of the four oils in the absence of any PDA particles led to no/unstable emulsions: rapid macro-phase separation occurred. These results indicate that the PDA particles play an important role for the stabilization of the emulsions.

3.2.2. Effects of PDA Particles Concentration on Emulsion Formation and Stability

It is worth asking whether the PDA particle concentration can be reduced below 1.00 wt % without affecting the emulsifier performance. Table 1 summarizes the results obtained by systematically reducing the PDA particle concentration from 1.00 to 0.05 wt % for *n*-dodecane as model oil. There is a clear trend of increasing droplet size with a decrease of PDA particle concentration. All the emulsions obtained were confirmed to be oil-in-water type from drop test although poor emulsions were obtained at or below PDA particles concentration of 0.05 wt %. The percentage of survived emulsion was calculated referring an equation (Equation (1)).

$$Survived\,emulsion\,\% \; = \; 100 \, - \, \left(\frac{V_{oil}^{separated}}{V_{oil}^{initial}} \, \times \, 100 \right) \tag{1}$$

$V_{oil}{}^{initial}$: Initial oil volume prior to emulsification
$V_{oil}{}^{separated}$: Volume of separated oil

Table 1. Characterization data obtained for the emulsions prepared by adding PDA aqueous dispersion at various concentrations to *n*-dodecane. Equal volumes of oil and aqueous PDA dispersion were used and emulsification was carried out at 20,000 rpm for 2 min.

PDA Concentration/wt %	Type of Emulsion Formed	Survived Emulsion for 1 Week/ %	Volume-Average oil droplet Diameter/ μm
0.05	Oil/water	81	88 ± 40
0.10	Oil/water	99.5	46 ± 25
0.20	Oil/water	99.5	44 ± 24
0.50	Oil/water	~100	33 ± 13
1.00	Oil/water	~100	30 ± 12

At and above the PDA particles concentration of 0.50 wt %, the emulsions were completely stable to coalescence over one week and only slow creaming was observed with time (see Figure 2a). At the PDA particle concentrations of 0.20 and 0.10 wt %, the emulsions were relatively stable: 100% emulsion survived for 24 h and only approximately 0.5% demulsification occurred after one week. At the PDA concentration of 0.05 wt %, the emulsion survived well (81% after one week), however demulsification occurred slowly to result in macrophase separation (only 66% emulsion survived) after 1.5 months. Typical OM images taken 24 h after emulsification in Figure 3 showed polydisperse oil droplets prepared at every PDA concentration. Number-average diameters were estimated using the optical micrographs as follows: 1.00 wt %, 11 ± 6 µm; 0.50 wt %, 12 ± 3 µm; 0.20 wt %, 13 ± 4 µm; 0.10 wt %, 17 ± 4 µm; and 0.05 wt %, 51 ± 14 µm (*n* =100). These number-average oil droplet diameters increased with a decrease of the PDA particle concentration, whose tendency accorded well with that estimated by the laser diffraction method (see Figure 2b and Table 1).

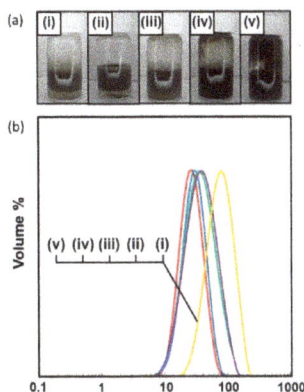

Figure 2. (**a**) Digital photographs; and (**b**) droplet size distribution curves of PDA-stabilized *n*-dodecane-in-water emulsions prepared at various PDA concentrations: (i) 0.05 wt %; (ii) 0.10 wt %; (iii) 0.20 wt %; (iv) 0.50 wt %; and (v) 1.00 wt %.

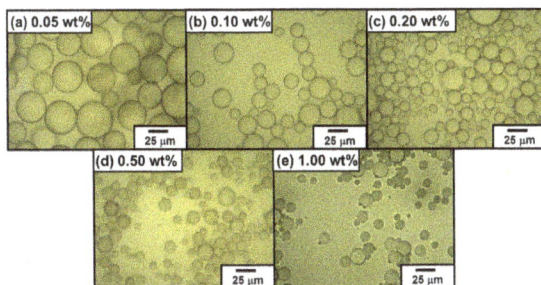

Figure 3. Optical micrographs of PDA-stabilized "Pickering-type" *n*-dodecane-in-water emulsions prepared at various PDA concentrations: (**a**) 0.05 wt %; (**b**) 0.10 wt %; (**c**) 0.20 wt %; (**d**) 0.50 wt %; and (**e**) 1.00 wt %. The emulsions were observed 24 h after preparation.

The fraction of PDA particles adsorbed at the oil–water interface can be readily estimated assuming a monolayer of adsorbed PDA particles is formed. Percentages of the PDA particles effectively attached on the oil–water interface based on the total amount of the PDA particles added were calculated using a following simple equation (Equation (2)) [17].

$$\% \, PDA \, particles = \pi \frac{R_{oil}^2 \, N_{oil}}{N_{part} \, R_{part}^2} \times 100 \, \% \tag{2}$$

where $N_{part} = \dfrac{W_{part} \, N_A}{Mw_{part}}$, and $N_{oil} = \dfrac{3 \, V_{oil}}{4 \, \pi \, R_{oil}^3}$.

Here, the calculations require some assumptions regarding the PDA particles packing efficiency at the interface, and the relatively polydisperse nature of the droplets can also introduce errors [13]. We assume 2D square lateral packing, uniform PDA particles and droplet sizes and PDA particle dimensions negligible as compared with those of oil droplets. We also assume that there are no PDA particles present in the oil phase because the energy barrier for the PDA particles to enter *n*-dodecane phase is too high. R_{part} and R_{oil} are the radii of the particles and oil, respectively (R_{oil} values used here were the volume mean radius determined by the laser diffraction method); N_{part} and N_{oil} are the numbers of particles and oil droplets, respectively; V_{oil} is the volume of oil; and W_{part} is the weight of PDA particles. The density of PDA (1.52 g/cm^3) was used to calculate the PDA particle numbers used for preparation of emulsions. Laser diffraction studies (see Table 1) suggest mean volume-average droplet diameters of 30 ± 12, 33 ± 13, 44 ± 24, 46 ± 25 and 88 ± 40 μm obtained at the PDA particle concentrations of 1.00, 0.50, 0.20, 0.10 and 0.05 wt %, respectively. Within these constraints, the PDA particle adsorption efficiencies were estimated to be 352%, 682%, 1207%, 2307% and 1957%. This is a surprising result, and we need to re-examine the assumptions used for the calculations. The possible reason for these values over 100% efficiency should be due to non-closely packed PDA particles at oil–water interface in aqueous media at all the PDA particle concentrations. Recently, it has been reported that the PDA particles are expected to consist of electrostatically bonded PDA and DA oligomers [42,50,51], and there is a possibility that these DA oligomers can be dissolved into aqueous medium during storage after extensive centrifugal washing and could work as a molecular emulsifier. Actually, homogenization of the supernatant of the PDA particle aqueous dispersion, which was prepared by centrifugation, and *n*-dodecane led to formation of oil-in-water emulsion. From these results, it is expected that mixture of the PDA solid particles and DA oligomers eluting out from the PDA particles work as an emulsifier and the PDA particles are adsorbed to oil droplet surface in non-closely packed manner.

In order to investigate the particle adsorption density at oil droplet surface in detail, wax was used as a solidifiable oil phase to stabilize oil-in-water emulsion. The particle-stabilized emulsion was successfully prepared using wax (3.0 mL) with a melting point of 58–60 °C by homogenization with an aqueous dispersion of PDA particles (3.0 mL, 0.50 wt %) at 70 °C. The prepared oil-in-water emulsion was cooled down to room temperature, and the liquid oil phase was solidified, which made possible for the droplets to be observed by SEM as well as OM (Figure 4). The droplet diameters were estimated to range from 10 to 200 μm (Heywood diameter, 123 ± 39 μm), and the PDA particles can be observed at droplet surface by OM (Figure 4a,b), thanks to their black color. Interestingly, the oil droplets were not fully covered with the PDA particles, and the PDA particles adsorbed at the surface formed islands. SEM studies also supported these OM results and magnified SEM images revealed the islands consisted of the PDA particle monolayer (Figure 4c,d). The area with no PDA particles should be stabilized with DA oligomers which cannot be observed using SEM due to their small sizes. There is a possibility that lateral capillary forces working between PDA particles at oil–water interface effectively attracted each other and the particles gathered to form islands [52]. Average percentage of the PDA particles at wax droplet surface was measured to be 42% ± 17%.

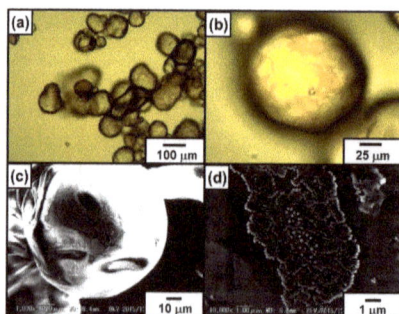

Figure 4. (**a,b**) Optical micrographs; and (**c,d**) SEM images of PDA-stabilized wax-in-water emulsion prepared using PDA particle aqueous dispersion (0.50 wt %). (**b,d**) are magnified images of (**a,c**), respectively. SEM images were taken without Au coating at an electron acceleration voltage of 8 kV.

3.2.3. Visualization of Transparent Emulsion

PDA is known to be deeply colored and this color can be easily monitored with the naked eye. Therefore, PDA has been considered to be a good candidate as a coloring agent for creating patterned surface [53]. In the present study, the performance of the PDA particles as a colored particulate emulsifier was evaluated. Properties of the PDA material used in this study are its intense, intrinsic chromogenicity and its colloidal dimensions. A mixture of 62 wt % perfluorotoluene and 38 wt % perfluorononane, which has the same refractive index as water, was used as an oil phase. The refractive index of the oil and water was matched; therefore, in this emulsion system, light is not refracted or reflected by the oil–water interface, which generally leads to transparent emulsions. The PDA particles (1.00 wt %) proved to act as an effective particulate emulsifier for the oil, and the droplet test for the emulsion indicated that an oil-in-water emulsion was obtained. A typical OM image of the emulsion is shown in Figure 5a. The colored PDA particles attached to the oil–water interface for visualization of the emulsion droplets. The emulsion droplets were spherical and fairly polydisperse, with diameters ranging from 70 to 250 µm. Careful OM studies confirmed that there were bare oil–water interfaces that were not covered with the PDA particles on the droplet surface (Figure 5b), as already observed in wax-in-water emulsion system. In contrast, droplets could hardly be observed for the emulsion stabilized with PVA, which acts as a surface-active polymeric stabilizer (Figure 5c). The diameters of the droplets stabilized with PVA were between 5 and 20 µm and were smaller than those of the droplets stabilized with PDA particles. This is because molecular-level PVA emulsifier could stabilize larger oil–water interfacial area comparing with particulate PDA emulsifier.

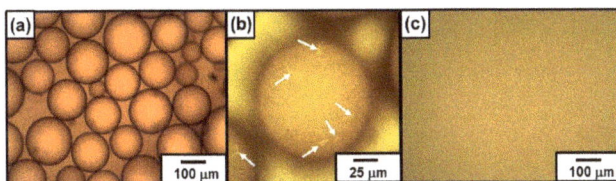

Figure 5. Optical micrographs of oil (mixture of 62% perfluorotoluene and 38 wt % perfluorononane)-in-water emulsion stabilized with (**a**,**b**) PDA particles and (**c**) PVA stabilizer. (**b**) is a magnified image of (**a**). The emulsion was prepared at 1.00 wt % stabilizer concentration and were observed 1 h after preparation. Arrows in (**b**) indicate bare oil–water interface that was not coated with the PDA particles.

3.2.4. Emulsion Data with Dichloromethane as a Model Volatile Oil

Recently, fabrication of colloidal assembly consisting of PDA particles has attained notable interest, because of their unique structural coloring character [49]. However, simple routes to direct and assemble PDA particles into shape-controlled constructs with hierarchical ordering are still lacking. We utilized the oil–water interface (that is, on the surface of oil droplets dispersed in continuous aqueous phase), which has been shown to be an ideal place for the assembly of colloidal particles, to fabricate capsule consisting of colloidal assembly shell. Homogenization of DCM and PDA aqueous dispersion (1.00 wt %) successfully led to stable DCM-in-water emulsion. OM studies recorded during the *in situ* evaporation of DCM gave important information of structure of the particle-stabilized oil droplets. The oil droplets, first, shrunk due to a decrease of oil volume remaining spherical shape. Then, slow droplet deformation from sphere to non-sphere occurred, and wrinkles appeared on the surface of droplet (see Figure 6). Relationship between time and circularity of oil droplets was shown in Figure 7, and it is clear that the circularity remained 1.0 until 4–5 min and then started to decrease. As indicated, the PDA particles were adsorbed to oil droplet surface in patchy manner, and it is expected that the droplet shrunk maintaining spherical shape until the PDA particles completely cover the droplet surface. After close-packed covering of the droplet surface with the PDA particles, the droplet shape started to deviate from sphere to non-sphere, because the volume of droplet decrease with keeping the fixed surface area. The appearance of the wrinkles is strong evidence for the attachment of PDA particles on the oil–water interface and stabilization of the emulsion. The SEM studies of the PDA residues remaining after evaporation of DCM and water from the emulsion revealed a wrinkled and ruptured capsule morphology (Figure 8a,b): this morphology accords well with that observed in the OM studies after the evaporation of DCM. The detailed SEM observation of the capsule confirmed the existence of cracks on the capsule surface, which should indicate there is no chemical bonding among the PDA particles and they attract each other via van der Waals force-based particle–particle interaction. Sonication of the PDA capsules dispersed in water medium led to breakage and formation of ill-defined PDA particle debris, which should indicate that the capsules were again formed via van der Waals force-based particle–particle interaction.

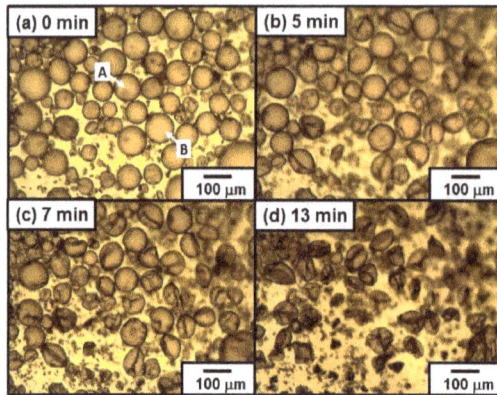

Figure 6. Optical micrographs illustrating evaporation of oil phase from DCM-in-water emulsion prepared using PDA aqueous dispersion (1.00 wt %). The emulsion was stored for 10 days after preparation before observation. The images were taken: (**a**) 0 min; (**b**) 5 min; (**c**) 7 min; and (**d**) 13 min after start of optical microscopy observation. The oil droplets indicated using arrows were used for estimation of circularity (see Figure 7).

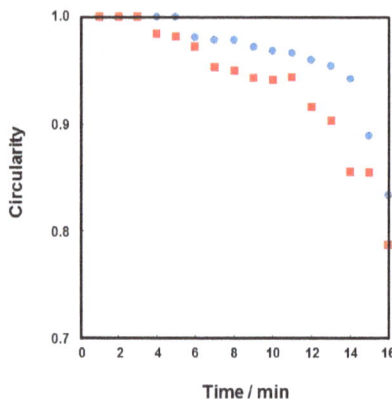

Figure 7. Relationship between time and circularity obtained for the oil droplets observed in Figure 6 (as indicated using arrows: ●, droplet A; ■, droplet B).

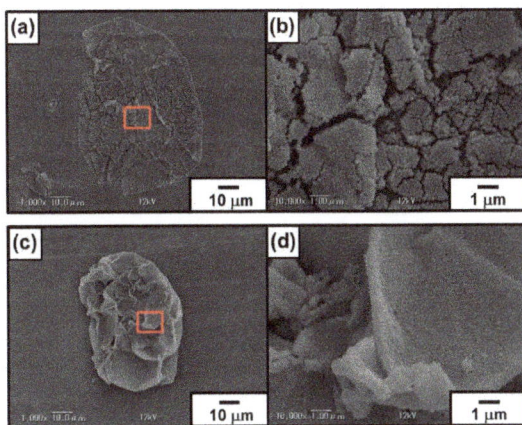

Figure 8. SEM images of "Pickering-type" DCM-in-water emulsion (**a,b**) without and (**c,d**) with crosslinking after the evaporation of DCM. (**b,d**) are magnified images of the areas shown in (**a,c**), respectively.

3.3. Formation of Colloidosomes from Particle-Stabilized Emulsion

In order to fabricate robust microcapsules, the PDA particles were connected with each other at oil–water interface of the droplets. These kinds of microcapsules are known as colloidosomes [20,23]. Covalent crosslinking is one of effective routes to connect the particles at oil–water interfaces, and some reactions (e.g., reactions between maleic anhydride and amine groups [54], epoxy and amine groups [55–58] and hydroxyl and isocyanate groups [59]) have been utilized. Here, the reaction between catechol and amine groups [44] was utilized to crosslink the PDA particles at droplet surfaces. Specifically, the PDA particles carrying catechol groups on their surfaces were crosslinked using water-soluble PEI crosslinker at oil–water interface. Although the contact angle of the PDA particles at oil–water interface is not known, the formation of an oil-in-water emulsion means that the contact angle must be less than 90° [3–5]. Thus, more than 50% surface area of each adsorbed particle should be exposed to the aqueous phase compared to the oil phase, which means large amount of the catechol groups on the particle surface must be available for reaction with the water-soluble PEI crosslinker.

Figure 9 shows OM images captured during evaporation of DCM from the colloidosomes prepared using 10 wt % PEI aqueous solution. The OM studies confirmed that the droplets had wrinkles on their surfaces from start of crosslinking reaction, rather than near-spherical shape observed in non-crosslinked precursor Pickering emulsion system. This difference should be due to partial removal of DCM oil phase from the droplets and decrease of the droplet surface area before crosslinking, which leads to close-packed PDA particles at the interface. Crosslinking reaction was conducted using the DCM-in-water emulsion with 10 days storage period at room temperature after preparation, and it is expected that small amount of DCM evaporated even closed with a lid, which should lead to near-closed packing of the PDA particles at droplet surface. The emulsion was diluted with PEI aqueous solution when the crosslinking reaction was conducted, which should also decrease the surface area of the droplets and close packing of PDA particles occur because of partial dissolution of DCM from the droplets into continuous aqueous phase.

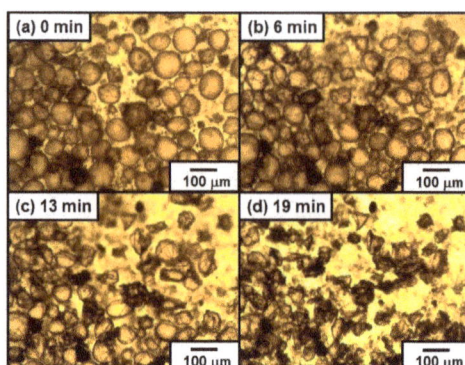

Figure 9. Optical micrographs illustrating evaporation of oil phase from colloidosomes prepared by crosslinking DCM-in-water emulsion shown in Figure 6. The images were taken: (**a**) 0 min; (**b**) 6 min; (**c**) 13 min; and (**d**) 19 min after start of optical microscopy observation.

Degree of circularity of colloidosome decreased with an increase of DCM evaporation time, as observed in precursor Pickering emulsion system. The SEM studies of the colloidosomes after evaporation of DCM and water revealed a wrinkled capsule morphology (Figure 8c,d). The microcapsule with wrinkles had few cracks on their surface, which should indicate there is crosslinking of the PDA particles successfully occurred. Successful crosslinking was also assessed by a sonication challenge, followed by OM and SEM observations. Sonication of the colloidosomes dispersed in aqueous medium after removal of DCM retained their microcapsule morphologies even after extensive sonication (Figure 10f–h), which should again indicate that the successful crosslinking occurred. On the other hand, the residues obtained after oil phase removal from the PDA particle-stabilized emulsion prepared in the absence of any crosslinker were broken into small fragments of PDA particle flocs after sonication (Figure 10b–d).

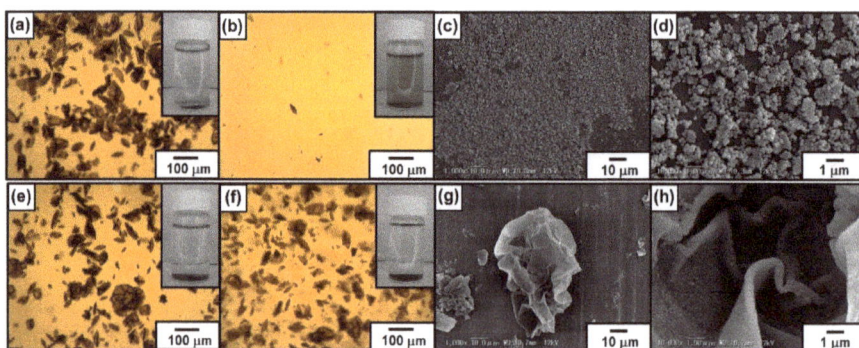

Figure 10. (**a,b,e,f**) Optical micrographs and (**c,d,g,h**) SEM images of PDA particle-stabilized DCM-in-water emulsions (**a–d**) without and (**e–h**) with crosslinking using 10 wt % PEI aqueous solution after removal of DCM oil phase. The emulsions were observed (**a,e**) before and (**b–d,f–h**) after sonication challenges. Insets shown in (**a,b,e,f**) are digital photographs of the samples.

It is worth investigating how much the PEI crosslinker concentration can be reduced without affecting colloidosome formation (Figure 11). OM studies on colloidosomes after extensive sonication challenge revealed that wrinkled microcapsules were main product for 5 wt % PEI solution system. At and below 1 wt % PEI solution systems, the broken capsules were observed and the amount of ill-defined PDA particle debris increased with a decrease of PEI concentration. These results indicated that the robustness of colloidosomes can be controlled by changing the crosslinking degree.

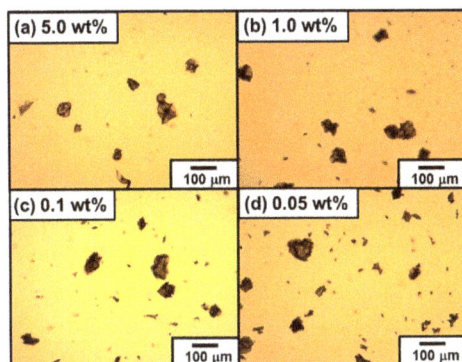

Figure 11. Optical micrographs of PDA particle-stabilized DCM-in-water emulsions after crosslinking using PEI aqueous solutions with various concentrations: (**a**) 5.0 wt %; (**b**) 1.0 wt %; (**c**) 0.1 wt %; and (**d**) 0.05 wt % PEI aqueous solution systems. The emulsions after removal of DCM oil phase were observed after sonication challenges.

4. Conclusions

In summary, we described the first use of PDA as a particulate emulsifier. The PDA particles proved to be an effective Pickering emulsifier for the stabilization of oil-in-water emulsions. These emulsions were characterized in terms of their mean droplet diameter and morphology using laser diffraction, OM and SEM. SEM studies of the PDA residues remaining after evaporation of oil and water from the emulsion revealed a capsule morphology, which is strong evidence for the attachment of PDA particles on the oil–water interface and stabilization of the emulsion. The PDA particles adsorbed to the oil–water interface can be covalently crosslinked using PEI to form colloidosomes, which can retain their microcapsule morphology against sonication challenge. The understanding and thereafter the control over the unique organizations of the particle assembly at fluid–fluid interfaces render the possibility to fabricate functional nanostructured materials with hierarchical orderings, like ultrathin particle membranes. Chemical functionalization of PDA can open further opportunities for their use in different applications. These new colloidosomes are expected to become a useful drug delivery carrier, catalyst, and cosmetics.

Acknowledgments: This work was partially supported by a Grant-in-Aid for Scientific Research on Innovative Areas "Engineering Neo-Biomimetics (No. 15H01602 and 15H01593)", "New Polymeric Materials Based on Element-Blocks (No. 15H00767)", and "Molecular Soft Interface Science (No. 23106720)" from the Ministry of Education, Culture, Sports, Science, and Technology of Japan.

Author Contributions: Nobuaki Nishizawa carried out the experiments with respect to preparation and characterization of PDA particle-stabilized emulsions. Ayaka Kawamura carried out synthesis and characterization of PDA particles. Yoshinobu Nakamura carried out characterization of PDA particle-stabilized emulsions. Syuji Fujii organized the project and had the idea for the PDA particle-stabilized Pickering emulsion. Syuji Fujii and Michinari Kohri wrote the manuscript. All authors discussed the results and edited the manuscript.

Conflicts of Interest: The authors declare no conflict of interest.

Abbreviations

PDA	Polydopamine
DA–HCl	Dopamine hydrochloride
DCM	Dichloromethane

PEI   Poly(ethylene imine)
PVA   Poly(vinyl alcohol)
SEM   Scanning electron microscopy
DLS   Dynamic light scattering
OM   Optical microscopy

References

1. Ramsden, W. Separation of solids in the surface-layers of solutions and "suspensions" (observations on surface-membranes, bubbles, emulsions, and mechanical coagulation)-Preliminary account. *Proc. R. Soc. Lond.* **1903**, *72*, 156–164. [CrossRef]
2. Pickering, S.U. CXCVI.-emulsions. *J. Chem. Soc. Trans.* **1907**, *91*, 2001–2021. [CrossRef]
3. Binks, B.P.; Horozov, T.S. *Colloidal Particles at Liquid Interfaces*; Cambridge University Press: Cambridge, UK, 2006.
4. Binks, B.P. Particles as surfactants-similarities and differences. *Curr. Opin. Colloid Interface Sci.* **2002**, *7*, 21–41. [CrossRef]
5. Aveyard, R.; Binks, B.P.; Clint, J.H. Emulsions stabilised solely by colloidal particles. *Adv. Colloid Interface Sci.* **2003**, *100–102*, 503–546. [CrossRef]
6. Binks, B.P.; Lumsdon, S.O. Stability of oil-in-water emulsions stabilised by silica particles. *Phys. Chem. Chem. Phys.* **1999**, *1*, 3007–3016. [CrossRef]
7. Binks, B.P.; Lumsdon, S.O. Catastrophic phase inversion of water-in-oil emulsions stabilized by hydrophobic silica. *Langmuir* **2000**, *16*, 2539–2547. [CrossRef]
8. Duan, H.; Wang, D.; Kurth, D.G.; Möhwald, H. Directing self-assembly of nanoparticles at water/oil interface. *Angew. Chem. Int. Ed.* **2004**, *43*, 5639–5642. [CrossRef] [PubMed]
9. Duan, H.; Wang, D.; Sobal, N.S.; Giersig, M.; Kurth, D.G.; Möhwald, H. Magnetic colloidosomes derived from nanoparticle interfacial self-assembly. *Nano Let.* **2005**, *5*, 949–952.
10. Wang, D.; Duan, H.; Möhwald, H. The water/oil interface: The emerging horizon for self-assembly of nanoparticles. *Soft Matter.* **2005**, *1*, 412–416. [CrossRef]
11. Lin, Y.; Skaff, H.; Böker, A.; Dinsmore, A.D.; Emrick, T.; Russell, T.P. Ultrathin cross-linked nanoparticle membranes. *J. Am. Chem. Soc.* **2003**, *125*, 12690–12691. [CrossRef] [PubMed]
12. Lin, Y.; Skaff, H.; Emrick, T.; Dinsmore, A.D.; Russell, T.P. Nanoparticle assembly and transport at liquid-liquid interfaces. *Science* **2003**, *299*, 226–229. [CrossRef] [PubMed]
13. Cauvin, S.; Colver, P.J.; Bon, S.A.F. Pickering stabilized miniemulsion polymerization: Preparation of clay armored latexes. *Macromolecules* **2005**, *38*, 7887–7889. [CrossRef]
14. Fujii, S.; Okada, M.; Furuzono, T. Hydroxyapatite nanoparticles as stimuls-responsive particulate emulsifiers and building block for porous materials. *J. Colloid Int. Sci.* **2007**, *315*, 287–296. [CrossRef] [PubMed]
15. Fujii, S.; Okada, M.; Sawa, H.; Furuzono, T.; Nakamura, Y. Hydroxyapatite nanoparticles as particulate emulsifier: Fabrication of hydroxyapatite-coated biodegradable microspheres. *Langmuir* **2009**, *25*, 9759–9766. [CrossRef] [PubMed]
16. Kaur, G.; He, J.; Xu, J.; Pingali, S.; Jutz, G.; Böker, A.; Niu, Z.; Li, T.; Rawlinson, D.; Emrick, T.; Lee, B.; Thiyagarajamn, P.; Russell, T.P.; Wang, Q. Interfacial assembly of turnip yellow mosaic virus nanoparticles. *Langmuir* **2009**, *25*, 5168–5176. [CrossRef] [PubMed]
17. Fujii, S.; Aichi, A.; Muraoka, M.; Kishioto, N.; Iwahori, K.; Nakamura, Y.; Yamashita, I. Ferritin as a bionano-particulate emulsifier. *J. Colloid Interface Sci.* **2009**, *338*, 222–228. [CrossRef] [PubMed]
18. Van Rijn, P.; Mougin, N.C.; Franke, D.; Park, H.; Böker, A. Pickring emulsion templated soft capsules by self-assembling cross-linkable ferritin-polymer conjugates. *Chem. Commun.* **2011**, *47*, 8376–8371. [CrossRef] [PubMed]
19. Schulz, A.; Liebeck, B.M.; John, D.; Heiss, A.; Subkowskic, T.; Böker, A. Protein–mineral hybrid capsules from emulsions stabilized with an amphiphilic protein. *J. Mater. Chem.* **2011**, *21*, 9731–9736. [CrossRef]
20. Velev, O.D.; Furusawa, K.; Nagayama, K. Assembly of latex particles by using emulsion droplets as templates. 1. Microstructured hollow spheres. *Langmuir* **1996**, *12*, 2374–2384. [CrossRef]
21. Cayre, O.J.; Noble, P.F.; Paumov, V.N. Fabrication of novel colloidsome microcapsules with gelled aqueous cores. *J. Mater. Chem.* **2004**, *14*, 3351–3355. [CrossRef]

22. Binks, B.P.; Rodrigues, J.A. Inversion of emulsions stabilized solely by ionizable nanoparticles. *Angew. Chem. Int. Ed.* **2005**, *44*, 441–444. [CrossRef] [PubMed]

23. Dinsmore, A.D.; Hsu, M.F.; Nikolaides, M.G.; Marquez, M.; Bausch, A.R.; Weitz, D.A. Colloidsomes: Selectively permeable capsules composed of colloidal particles. *Science* **2002**, *298*, 1006–1009. [CrossRef] [PubMed]

24. Amalvy, J.I.; Armes, S.P.; Binks, B.P.; Rodrigues, J.A.; Unali, G.-F. Use of sterically-stabilised polystyrene latex particles as a pH-responsive particulate emulsifier to prepare surfactant-free oil-in-water emulsions. *Chem. Commun.* **2003**, *15*, 1826–1827. [CrossRef]

25. Binks, B.P.; Murakami, R.; Armes, S.P.; Fujii, S. Temperature-induced inversion of nanoparticle-stabilized emulsions. *Angew. Chem. Int. Ed.* **2005**, *44*, 4795–4798. [CrossRef] [PubMed]

26. Fujii, S.; Randall, D.P.; Armes, S.P. Synthesis of polystyrene/poly[2-(dimethylamino)ethyl methacrylate-*stat*-ethylene glycol dimethacrylate] core-shell latex particles by seeded emulsion polymerization and their application as stimulus-responsive particulate emulsifiers for oil-in-water emulsions. *Langmuir* **2004**, *20*, 11329–11335.

27. Fujii, S.; Aichi, A.; Akamatsu, K.; Nawafune, H.; Nakamura, Y. One-step synthesis of polypyrrole-coated silver nanocomposite particles and their application as a coloured particulate emulsifier. *J. Mater. Chem.* **2007**, *17*, 3777–3779. [CrossRef]

28. Fujii, S.; Read, E.S.; Armes, S.P.; Binks, B.P. Stimulus-responsive emulsifiers based on nanocomposite microgel particles. *Adv. Mater.* **2005**, *17*, 1014–1018. [CrossRef]

29. Binks, B.P.; Murakami, R.; Armes, S.P.; Fujii, S. Effects of pH and salt concentration on oil-in-water emulsions stabilized solely by nanocomposite microgel particles. *Langmuir* **2006**, *22*, 2050–2057. [CrossRef] [PubMed]

30. Fujii, S.; Armes, S.P.; Binks, B.P.; Murakami, R. Stimulus-responsive particulate emulsifiers based on lightly cross-linked poly(4-vinylpyridine)-silica nanocomposite microgels. *Langmuir* **2006**, *22*, 6815–6825. [CrossRef] [PubMed]

31. Ngai, T.; Behrens, S.H.; Auweter, H. Novel emulsions stabilized by pH and temperature sensitive microgels. *Chem. Commun.* **2005**, *3*, 331–333. [CrossRef] [PubMed]

32. Fujii, S.; Cai, Y.; Weaver, J.V.M.; Armes, S.P. Syntheses of shell cross-linked micelles using acidic ABC triblock copolymers and their application as pH-responsive particulate emulsifiers. *J. Am. Chem. Soc.* **2005**, *127*, 7304–7305. [CrossRef] [PubMed]

33. Lee, H.; Dellatore, S.M.; Miller, W.M.; Messersmith, P.B. Mussel-inspired surface chemistry for multifunctional coatings. *Science* **2007**, *318*, 426–430. [CrossRef] [PubMed]

34. Cui, J.; Yan, Y.; Such, G.K.; Liang, K.; Ochs, C.J.; Postma, A.; Caruso, F. Immobilization and intracellular delivery of an anticancer drug using mussel-inspired polydopamine capsules. *Biomolecules* **2012**, *13*, 2225–2228. [CrossRef] [PubMed]

35. Liu, R.; Guo, Y.L.; Odusote, G.; Qu, F.L.; Priestley, R.D. Core–shell Fe$_3$O$_4$ polydopamine nanoparticles serve multipurpose drug carrier, catalyst support and carbon adsorbent. *ACS Appl. Mater. Interfaces* **2013**, *5*, 9167–9171. [CrossRef] [PubMed]

36. Gao, H.C.; Sun, Y.M.; Zhou, J.J.; Xu, R.; Duan, H.W. Mussel-inspired synthesis of polydopamine-functionalized graphene hydrogel as reusable adsorbents for water purification. *ACS Appl. Mater. Interfaces* **2013**, *5*, 425–432. [CrossRef] [PubMed]

37. Fu, J.; Chen, Z.; Wang, M.; Liu, S.; Zhang, J.; Zhang, J.; Han, R.; Xu, Q. Adsorption of methylene blue by a high-efficiency adsorbent (polydopamine microspheres): Kinetics, isotherm, thermodynamics and mechanism analysis. *Chem. Eng. J.* **2015**, *259*, 53–61. [CrossRef]

38. Kohri, M.; Kohma, H.; Shinoda, Y.; Yamauchi, M.; Yagai, S.; Kojima, T.; Taniguchi, T.; Kishikawa, K. A colorless functional polydopamine thin layer as a basis for polymer capsules. *Polym. Chem.* **2013**, *4*, 2696–2702. [CrossRef]

39. Kohri, M.; Shinoda, Y.; Kohma, H.; Nannichi, Y.; Yamauchi, M.; Yagai, S.; Kojima, T.; Taniguchi, T.; Kishikawa, K. Facile synthesis of free-standing polymer brush films based on a colorless polydopamine thin layer. *Macromol. Rapid Commun.* **2013**, *34*, 1220–1224. [CrossRef] [PubMed]

40. Kohma, H.; Uradokoro, K.; Kohri, M.; Taniguchi, T.; Kishikawa, K. Hierarchically structured by colorless polydopamine thin layer and polymer brush layer. *Trans. Mater. Res. Soc. Jpn.* **2014**, *39*, 157–160. [CrossRef]

41. Kohri, M.; Kohma, H.; Uradokoro, K.; Taniguchi, T.; Kishikawa, K. Fabrication of colored particles covered by dye-bearing colorless polydopamine layer. *J. Colloid Sci. Biotechnol.* **2014**, *3*, 337–342.

42. Kohri, M.; Nannichi, Y.; Kohma, H.; Abe, D.; Kojima, T.; Taniguchi, T.; Kishikawa, K. Size control of polydopamine nodules formed on polystyrene particles during dopamine polymerization with carboxylic acid-containing compounds for the fabrication of raspberry-like particles. *Colloids Surf. A.* **2014**, *449*, 114–120. [CrossRef]

43. Kang, K.; Lee, S.; Kim, R.; Choi, I.S.; Nam, Y. Electrochemically driven, electrode-addressable formation of functionalized polydopamine films for neural interfaces. *Angew. Chem. Int. Ed.* **2012**, *51*, 13101–13104. [CrossRef] [PubMed]

44. Tian, Y.; Cao, Y.; Wang, Y.; Yang, W.; Feng, J. Realizing ultrahigh modulus and high strength of macroscopic graphene oxide papers through crosslinking of mussel-inspired polymers. *Adv. Mater.* **2013**, *25*, 2980–2983. [CrossRef] [PubMed]

45. Farnad, N.; Farhadi, K.; Voelcker, N.V. Polydopamine nanoparticles as a new and highly selective biosorbent for the removal of copper (II) ions from aqueous solutions. *Water Air Soli Pollut.* **2012**, *223*, 3535–3544. [CrossRef]

46. Ho, C.C.; Ding, S.J. The pH-controlled nanoparticles size of polydopamine for anti-cancer drug delivery. *J. Mater. Sci. Mater Med.* **2013**, *24*, 2381–2390. [CrossRef] [PubMed]

47. Yue, Q.; Wang, M.; Sun, Z.; Wang, C.; Wang, C.; Deng, Y.; Zhao, D. A versatile ethanol-mediated polymerization of dopamine for efficient surface modification and the construction of functional core-shell nanostructures. *J. Mater. Chem. B* **2013**, *1*, 6085–6093. [CrossRef]

48. Ai, K.; Liu, Y.; Ruan, C.; Lu, L.; Lu, G. Sp^2 C-dominant N-doped carbon sub-micrometer spheres with a tunable size: A versatile platform for highly efficient oxygen-reduction catalysts. *Adv. Mater.* **2013**, *25*, 998–1003. [CrossRef] [PubMed]

49. Kohri, M.; Nannichi, Y.; Taniguchi, T.; Kishikawa, K. Biomimetic non-iridescent structural color materials from polydopamine black particles that mimic melanin granules. *J. Mater. Chem. C* **2015**, *3*, 720–724. [CrossRef]

50. Hong, S.; Na, Y.S.; Choi, S.; Song, I.T.; Kim, W.Y.; Lee, H. Non-covalent self-assembly and covalent polymerization co-contribute to polydopamine formation. *Adv. Funct. Mater.* **2012**, *22*, 4711–4717. [CrossRef]

51. Liebscher, J.; Mrówczyński, R.; Scheidt, H.A.; Filip, C.; Hădade, N.D.; Turcu, R.; Bende, A.; Beck, S. Structure of polydopamine: A never-ending story? *Langmuir* **2013**, *29*, 10539–10548. [CrossRef] [PubMed]

52. Kralchevsky, P.A.; Nagayama, K. Capillary interactions between particles bound to interfaces, liquid films and biomembranes. *Adv. Colloid Interface Sci.* **2000**, *85*, 145–192. [CrossRef]

53. Zhang, L.; Yu, H.; Zhao, N.; Dang, Z.M.; Jian Xu, J. Patterned polymer surfaces with wetting contrast prepared by polydopamine modification. *J. Appl. Polym. Sci.* **2014**, *131*, 41057. [CrossRef]

54. Croll, L.M.; Stöver, H.D.H. Formation of tectocapsules by assembly and cross-linking of poly(divinylbenzene-*alt*-maleic anhydride) spheres at the oil-water interface. *Langmuir* **2003**, *19*, 5918–5922. [CrossRef]

55. Thompson, K.L.; Armes, S.P. From well-defined macromonomers to sterically-stabilised latexes to covalently cross-linkable colloidosomes: Exerting control over multiple length scales. *Chem. Commun.* **2010**, *46*, 5274–5276. [CrossRef] [PubMed]

56. Walsh, A.; Thompson, K.L.; Armes, S.P.; York, D.W. Polyamine-functional sterically stabilized latexes for covalently cross-linkable colloidosomes. *Langmuir* **2010**, *26*, 18039–18048. [CrossRef] [PubMed]

57. Williams, M.; Armes, S.P.; Verstraete, P.; Smets, J. Double emulsions and colloidosomes-in-colloidosomes using silica-based Pickering emulsifiers. *Langmuir* **2014**, *30*, 2703–2711. [CrossRef] [PubMed]

58. Cui, Y.; van Duijneveldt, J.S. Microcapsules composed of cross-linked organoclay. *Langmuir* **2012**, *28*, 1753–1757. [CrossRef] [PubMed]

59. Thompson, K.L.; Armes, S.P.; Howse, J.R.; Ebbens, S.; Ahmad, I.; Zaidi, J.H.; York, D.W.; Burdis, J.A. Covalently cross-linked colloidosomes. *Macromolecules* **2010**, *43*, 10466–10474. [CrossRef]

![polymers logo] *polymers*

MDPI

Article

Improved Stability of Emulsions in Preparation of Uniform Small-Sized Konjac Glucomanna (KGM) Microspheres with Epoxy-Based Polymer Membrane by Premix Membrane Emulsification

Yace Mi [1,2,†], Juan Li [1,†], Weiqing Zhou [1], Rongyue Zhang [1], Guanghui Ma [1,*] and Zhiguo Su [1]

[1] National Key Laboratory of Biochemical Engineering , Institute of Process Engineering,
    Chinese Academy of Sciences, Beijing 100190, China; miyacetx@163.com (Y.M.); lijuan@ipe.ac.cn (J.L.);
    wqzhou@ipe.ac.cn (W.Z.); ryzhang@iccas.ac.cn (R.Z.); zgsu@ipe.ac.cn (Z.S.)
[2] University of Chinese Academy of Sciences, Beijing 100049, China
*   Correspondence: ghma@ipe.ac.cn; Tel.: +86-10-8262-7072
†   These authors contribute equally to this work.

Academic Editor: Francoise M. Winnik
Received: 26 January 2016; Accepted: 16 February 2016; Published: 23 February 2016

Abstract: Uniform small-sized (<10 μm) Konjac glucomanna (KGM) microspheres have great application prospects in bio-separation, drug delivery and controlled release. Premix membrane emulsification is an effective method to prepare uniform small-sized KGM microspheres. However, since KGM solution bears strong alkalinity, it requires the membrane to have a hydrophobic surface resistant to alkali. In this study, uniform small-sized KGM microspheres were prepared with epoxy-based polymer membrane (EP) we developed by premix membrane emulsification. It was found that emulsion coalescence and flocculation occurred frequently due to the high interface energy and sedimentation velocity of KGM emulsions. Emulsion stability had a significant influence on the uniformity and dispersity of the final KGM microspheres. To improve the stability of the emulsions, the effects of the concentration of the emulsifier, the viscosity of the KGM solution, the oil phase composition and the feeding method of epoxy chloropropane (EC) on the preparation results were studied. Under optimal preparation conditions (emulsifier 5 wt % PO-5s, KGM III (145.6 mPa·s), weight ratio of liquid paraffin (LP) to petroleum ether (PE) 11:1), uniform and stable KGM emulsions (d = 7.47 μm, CV = 15.35%) were obtained and crosslinked without emulsion-instable phenomena.

Keywords: uniform small-sized KGM microspheres; premix membrane emulsification; epoxy-based polymer membrane (EP); alkaline condition

1. Introduction

Konjac glucomannan (KGM), as a high-molecular-weight polysaccharide, has special physical characteristics such as moisture retentivity, thickening, and gelling properties [1] and bears so many hydroxyl groups that various chemical modifications can be realized on KGM molecular chains [2–6]. KGM microspheres based on KGM powder have great application prospects in the fields of separation and purification [7], cell culture [8], enzyme and cell immobilization [9], drug delivery [10] and controlled release [11]. At present, the research on the preparation of KGM microspheres is insufficient, especially on those with small particle size (<10 μm). The most commonly used method to prepare KGM microspheres is mechanical stirring. In our previous work [12], a technology was developed to prepare KGM microspheres, which included acid degradation, alkali dissolution of KGM powder, emulsion preparation by mechanical stirring, and crosslinking steps. Compared with the traditional method, this method greatly shortened the reaction time and simplified the preparation process. However, since these microspheres prepared by this method were not uniform, sieving was necessary, and this resulted in higher product costs. It was more important to discover that uniform microspheres with small particle size (<10 μm) cannot be prepared by this method. Jianhua Shen [13] has reported a method of injection with template. It was characterized by the use of template to form homogeneous emulsions with controllable particle size. However, the size of the KGM microspheres prepared by this method was in the range of 50–500 μm, and small-sized (<10 μm) microspheres cannot be obtained.

Membrane emulsification technology [14], as a novel emulsion preparation method, has huge application prospects [15]. Different from the traditional technology, membrane emulsification can be used to produce uniform emulsions with a particle size less than 10 μm [16]. In this study, this technology was chosen to prepare small-sized KGM microspheres. KGM solutions used for the preparation of KGM microspheres bear strong alkalinity [17], which requires the membrane to have a hydrophobic surface resistant to alkali. However, Shirasu Porous Glass (SPG) membrane (Ise Chemical Co., Ise, Japan), as the most commonly used membrane in membrane emulsification [18], is chemically unstable in alkaline solution [19]. The main component of SPG is SiO_2 which is not resistant to alkali. If used in alkaline conditions chronically, SPG will become semitransparent, which means some pores in SPG have collapsed [20]. Furthermore, SPG is hydrophilic in nature due to a great amount of Si–OH groups on its surface [21]. Hydrophobic modification of SPG is needed in the preparation of water-in-oil (W/O) emulsions [22]. However, the hydrophobic ligands on the surface of SPG come off easily under alkaline conditions [19]. Therefore, SPG membrane cannot be directly used in the preparation of KGM microspheres. In a previous work, we obtained a kind of epoxy-based polymer (EP) membrane with a three-dimensional bicontinuous structure and stable hydrophobic surface [23]. With the appropriate porosity and narrow pore size distribution, as well as special pore structure, the EP membrane has been successfully applied in membrane emulsification to prepare W/O emulsions. It was worth noting that EP membrane can be applied in alkali systems due to its resistance to alkali [20]. In this study, uniform KGM emulsions were firstly prepared with EP membrane by premix membrane emulsification.

In the preparation process of KGM microspheres, emulsion coalescence and flocculation occur due to the high interface energy and sedimentation velocity of small-sized KGM emulsions [24]. Emulsion coalescence between two emulsion droplets is irreversible, and it will decrease the number and increase the size of emulsion droplets. Flocculation is a process where two or more emulsion droplets are adhered together into a whole by collision. These aggregated emulsion droplets will disperse again by shaking. However, if it occurs in the crosslinking process, those flocculated microspheres will be solidified and this flocculation is irreversible [25]. In practical cases, coalescence and flocculation may occur simultaneously. Some measures should be taken to avoid these phenomena and improve the preparation results.

In this study, the EP membrane we developed was firstly used in basic system to prepare uniform KGM microspheres by premix membrane emulsification. The effects of the concentration of emulsifiers, the viscosities of KGM solutions and the oil phase composition on the emulsification and crosslinking results were systematically studied. Meanwhile, different crosslinking results were obtained with the same formula by changing the feeding methods of epoxy chloropropane (EC).

2. Materials and Methods

2.1. Materials

Konjac powder (the content of KGM was approximately 65% (w/w)) was obtained from Hu-Bei Konson Konjac Gum Co. Ltd. (Hubei, China). Hexaglycerinpenta ester (PO-5s) purchased from Sakamoto Yakuhin Kogya Co. Ltd. (Osaka, Japan) was used as emulsifier. Liquid paraffin (LP) and petroleum ether (PE) were chosen as oil phase and were supplied by Sinopharm and Tianjin Jin Dong Tian Zheng Precision Chemical Reagent Factory (Tianjin, China), respectively. Epoxy chloropropane (EC) provided by Xilong Co. Ltd. (Guangdong, China) was used as crosslinking agent. Other reagents were of analytical purity and were purchased from Beijing Chemical Reagent Company (Beijing, China).

2.2. Preparation of KGM Solutions and Determination of the Viscosities of KGM Solutions and Oil Phases

Since 1% (w/w) KGM dissolved in water takes on a gelatin texture, it is too viscous to disperse in oil phase. Therefore, it is necessary to degrade the KGM molecules into small ones to improve its flowability. Here, KGM solutions with different viscosities were obtained by acid degradation. The preparation procedure was shown in Figure 1. A certain amount of hydrochloric acid (HCl, 0.5 M) was mixed with deionized water, and then the Konjac powder was added. With enough stirring, the mixture became jelly and then was heated at 115 °C for 55 min. After that, 45% (w/w) sodium hydroxide was added under vigorous stirring. Finally, 8% (w/w) KGM solutions with different viscosities were prepared (Table 1).

The viscosities of KGM solutions and oil phases were measured with a Brookfield viscometer (DV-II+Pro) using spindle 62 (50 rpm and 40 °C) and spindle 61 (50 rpm and 25, 40, 50, 60, 70, and 80 °C), respectively.

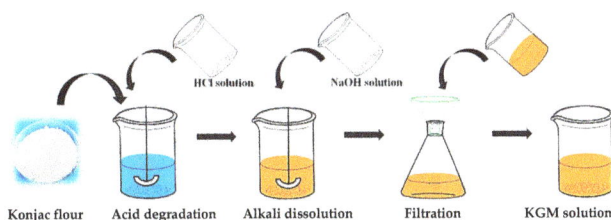

Figure 1. Scheme of the preparation of Konjac glucomanna (KGM) solution.

Table 1. Konjac glucomanna (KGM) solutions with different viscosities.

Samples	KGM I	KGM II	KGM III	KGM IV
Viscosity (mPa· s) [a]	66.0 ± 0.5	88.4 ± 0.5	145.6 ± 0.5	180.3 ± 0.5

[a] All of data were the average values from three determinations.

2.3. Preparation of KGM Microspheres by Premix Membrane Emulsification

The schematic diagram of premix membrane emulsification was shown in Figure 2. The coarse emulsions prepared by traditional method were pressed through a membrane with narrow pore size distribution under high pressure. The particle size of the obtained emulsions became smaller and more uniform. The tubular EP membrane used in this study has inside diameter of 0.8 cm, outside diameter of 1.0 cm and length of 2 cm.

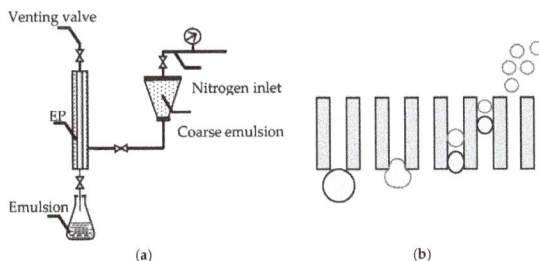

Figure 2. Schematic diagram of premix membrane emulsification apparatus and principle. (**a**) Premix membrane emulsification apparatus; (**b**) Premix membrane emulsification principle.

The preparation of KGM microspheres by premix membrane emulsification was as follows: the mixture of liquid paraffin and petroleum ether (weight ratio 7:5, 10:2, 11:1, 12:0) with an emulsifier PO-5s (3%, 4%, 5%, 6%) was chosen as oil phase. Aqueous phase of KGM solution (8.0 wt %, 3 mL) was emulsified in oil phase (40 mL) by homogenization (3600 rpm, 1 min) to form coarse emulsion. Then the coarse emulsion was poured into storage tank and pressed through EP membrane five times under nitrogen pressure to achieve uniform emulsions. EC (2 mL) as crosslinking agent was slowly added into the emulsions at 60 °C by burette, and then the mixtures were kept at 60 °C for 8 h to obtain the particles. Finally, the KGM particles were collected and successively washed by petroleum, ethanol and water. The pore sizes of EP membranes were 14.80 μm with operating pressure of 85 kPa. Recipe for the preparation of KGM microspheres was shown in Table 2.

Table 2. Recipe for preparing KGM microspheres.

Sample	Emulsifier Concentration in Oil Phase (*w/w*)	KGM Solution	Weight Ratio of Liquid Paraffin (LP) and Petroleum Ether (PE) (*w/w*)
S1	3%	KGM III	11:1
S2	4%	KGM III	11:1
S3	5%	KGM III	11:1
S4	6%	KGM III	11:1
S5	5%	KGM I	11:1
S6	5%	KGM II	11:1
S7	5%	KGM IV	11:1
S8	5%	KGM III	7:5
S9	5%	KGM III	10:2
S10	5%	KGM III	12:0

2.4. Determination of the Size and Size Distribution of KGM Emulsions

The average KGM emulsions size and size distribution coefficient of variation (*CV*) were determined by automatic analysis software on the basis of optical photographs taken by optical microscope (XSZ-H3). Diameters of over 500 droplets were analyzed to calculate the average particle size and particle size coefficient variation according to Equations (1) and (2). Wherein, d_i was the diameter of a single microsphere, $\bar{d}$ was the number-average particle diameter.

$$C.V. = \left(\sum_{i=1}^{n} \frac{\left(d_i - \bar{d}\right)^2}{N} \right)^{\frac{1}{2}} \bigg/ \bar{d} \tag{1}$$

$$\bar{d} = \sum_{i=1}^{n} d_i / N \tag{2}$$

2.5. Determination of Interfacial Tension between Aqueous and Oil Phases

The interfacial tension between aqueous and oil phases was determined by Contact angle system OCA (Dataphysics, Berlin, Germany). The aqueous phase was injected into the oil phase through the pinhole under the action of the microflow pump. This process would be recorded by video imaging system and then the interface tension was calculated through the video.

3. Results

In the preparation process of KGM microspheres with small particle size, emulsion coalescence and flocculation occurred due to the high interface energy and sedimentation velocity of KGM emulsions. Emulsion stability had a significant influence on the uniformity and dispersity of the final KGM microspheres. To improve the stability of emulsions, the effects of the concentration of the emulsifier, the viscosity of the KGM solution, the oil phase composition and the feeding method of EC on the preparation of KGM microspheres were studied.

3.1. Effects of Emulsifier Concentration on the Preparation of KGM Microspheres

Emulsifier adsorbed on the interface between oil and water could improve the stability of emulsions by forming a layer of viscoelastic membrane with a certain strength [26]. The emulsifier concentration has an important effect on the performance of membrane emulsification. After preliminary investigation, PO-5s was chosen as an emulsifier and the effects of its concentration on the preparation of KGM microspheres were investigated. Preparation parameters are shown in Table 2 (**S1**, **S2**, **S3**, **S4**).

The emulsification results are found in Figure 3 (**S1**, **S2**, **S3**, and **S4**). It shows that when the emulsifier concentration was increased from 3 wt % to 5 wt %, both the *CV* values and particle sizes of the KGM emulsions changed little. When the emulsifier concentration reached 6%, the *CV* value (20.64%) and the particle size (7.84 μm) had marked changes. The effects of the emulsifier concentration were not obvious until it was seriously excessive. The emulsifier played two main roles in the formation of the emulsions. One was lowering the interfacial tension between the oil and water, and the other one was stabilizing the emulsion droplets against coalescence and/or flocculation [27]. Increasing the amount of emulsifier to a proper amount could improve emulsion stability. However, when the emulsifier was excessive, it would play the opposite role [28]. The excess emulsifier will not further reduce the interfacial energy; instead, it will adsorb on the surface of the membrane pores. The more emulsifier was added, the more likely the EP membrane was covered with emulsifier following when it was wetted by the aqueous phase in the emulsification process. In this process, large-size droplets would be produced.

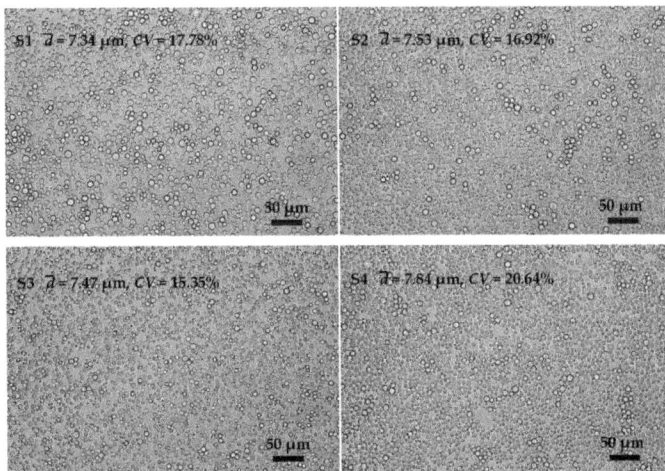

Figure 3. Microscopic photographs of KGM emulsions prepared with different emulsifier concentration (3%, 4%, 5% and 6%, Po-5S, *w*/*w*).

In the following crosslinking process of the KGM emulsions, the effects brought by the changes of emulsifier concentration cannot be ignored. The crosslinking results are shown in Figure 4 (S1-Crosslinking, S2-Crosslinking, S3-Crosslinking, and S4-Crosslinking). It was found that flocculation and coalescence occurred in S1-Crosslinking. Meanwhile, there were two sizes of particles which conglutinated to the gobbet. With the increase of emulsifier concentration, the phenomena of flocculation and coalescence were weakened and finally disappeared when the emulsifier concentration reached 5%. Emulsifier was the key factor in keeping the stability of the emulsions. Emulsions a with low content of emulsifier were unstable due to the high interface energy [29]. Increasing the emulsifier concentration would reduce the interfacial energy and improve the stability of the emulsions against the occurrence of flocculation and coalescence. In summary, the stability of the emulsions was important for the crosslinking results and 5% (w/w) was the optimal emulsifier concentration in terms of the uniformity and stability of emulsions.

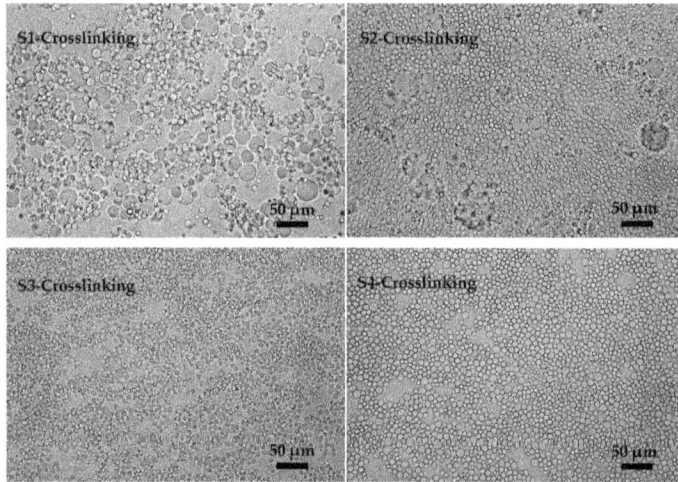

Figure 4. Microscopic photographs of KGM microspheres prepared with different emulsifier concentrations (3%, 4%, 5% and 6%, Po-5S, w/w): S1-Crosslinking, S2-Crosslinking, S3-Crosslinking and S4-Crosslinking were the crosslinking results of **S1**, **S2**, **S3** and **S4**.

3.2. Effects of the Viscosity of KGM Solution on the Preparation of KGM Microspheres

The content of KGM solution has great significance for the application of KGM microspheres. Since viscosities were the main difference in KGM solutions with different contents, the effects of the viscosity of KGM solution on the preparation of microspheres were studied. By changing the amount of acid in the process of acid degradation, four kinds of KGM solutions (Table 1) with different viscosities (66.0, 88.4, 145.6 and 180.3 mPa·s) were obtained.

KGM emulsions prepared with different viscosities of KGM solutions as aqueous phase are shown in Figure 5 (**S5**, **S6**, **S3**, and **S7**). The emulsification results varied little in the particle size and uniformity of emulsions, except for the emulsions prepared with a low viscosity of KGM solutions (Figure 5 (**S5**)), the uniformity of which was relatively poor ($CV = 22.55\%$). Coalescence could occur in the emulsification process due to the low viscosity of the aqueous phase, which would decrease the uniformity of the final emulsions. Therefore, high viscosity of aqueous phase showed advantages for obtaining uniform emulsions.

Figure 6 (S5-Crosslinking, S6-Crosslinking, S3-Crosslinking, and S7-Crosslinking) shows the crosslinking results. With a low viscosity of the KGM solution (Figure 6 (S5-Crosslinking)), the final microspheres had a much larger size compared with those before crosslinking. These large microspheres adhered to each other. By increasing the viscosity of KGM solution (Figure 6 (S6-Crosslinking)), two sizes of microspheres were produced. Compared with those in **S5**, the flocculation and coalescence were apparently weakened. The dispersity and uniformity of the final microspheres got much better after a further increase of the viscosity of the KGM solution (Figure 6 (S6-Crosslinking and S7-Crosslinking)). Low viscosity of the KGM solution possessed good liquidity contributing to mass transfer, which would induce coalescence in the collision process between emulsion droplets. In addition, low viscosity of KGM solution had low viscoelasticity, which would cause flocculation [30]. In the viscosity range we have studied, the KGM microspheres prepared with high viscosity of aqueous phases had good uniformity and dispersity. The greater the viscosity of the KGM solution, the higher the transmembrane pressure needed in premix membrane emulsification was. So considering the transmembrane pressure, we chose KGM III as the optimal aqueous phase.

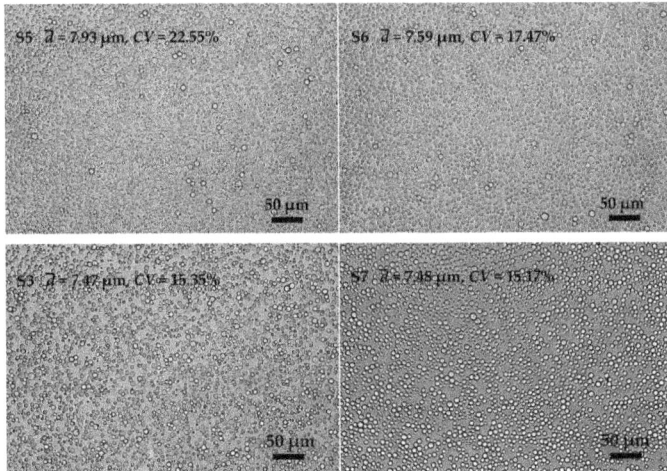

Figure 5. Microscopic photographs of KGM emulsions prepared with different KGM viscosities (KGM I, KGM II, KGM III and KGM IV).

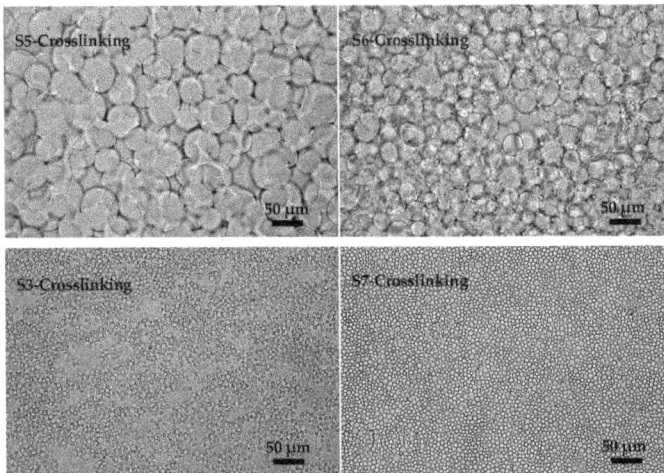

Figure 6. Microscopic photographs of KGM microspheres prepared with different KGM viscosities (KGM I, KGM II, KGM III and KGM IV): S5-Crosslinking, S6-Crosslinking, S3-Crosslinking and S7-Crosslinking were the crosslinking results of **S5**, **S6**, **S3** and **S7**.

3.3. Effects of Oil Phase Composition on the Preparation of KGM Microspheres

As one of the major components of emulsions, the continuous phase has a decisive influence on the stability of emulsions. We investigated the effects of oil phase composition in the following work. Figure 7 (**S8**, **S9**, **S3**, and **S10**) shows the preparation results of KGM emulsions by premix membrane emulsification. We found that the oil phase composition had little effect on the droplet size and size uniformity of the emulsions. However, the changes of the oil phase composition had considerable influence in the crosslinking process.

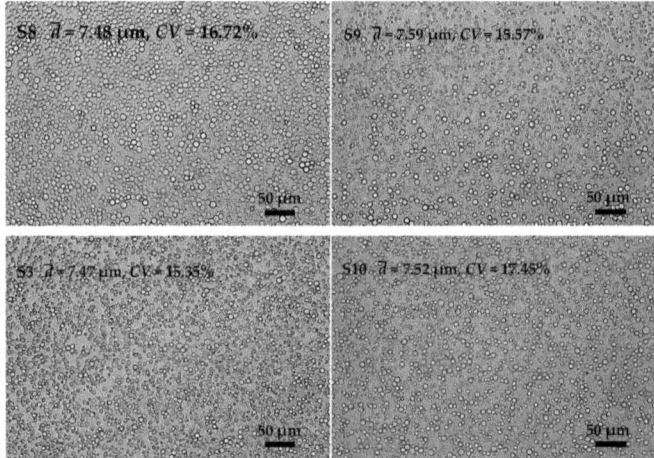

Figure 7. Microscopic photographs of KGM emulsions prepared with different oil phase compositions (7:5, 10:2, 11:1 and 12:0, LP/EP, *w*/*w*).

Since the viscosity of liquid paraffin was much higher than that of petroleum ether, the viscosity of the mixed oil phase decreased with the increase of the amount of petroleum ether (shown in Figure 8). In Figure 9 (S8-Crosslinking, S9-Crosslinking, S3-Crosslinking, and S10-Crosslinking), the crosslinking results with different oil phase composition are shown. Serious flocculation occurred in Figure 9 (S8-Crosslinking) due to the low viscosity of the oil phase. By increasing the viscosity of the oil phase, flocculation was weakened (Figure 9 (S9-Crosslinking)). As the viscosity of the oil phase was further increased, the dispersity of the final microspheres became better (Figure 9 (S3-Crosslinking and S10-Crosslinking)). When droplets approached each other, the discharging rate of the liquid between the deformed droplets partly depended on the viscosity of the continuous phase. The greater the viscosity was, the lower the liquid discharging rate and the flocculation rate were [31]. Considering the effects of oil phase composition on emulsification and crosslinking results, the optimum oil phase composition was 11:1 (*w*/*w*, LP/EP).

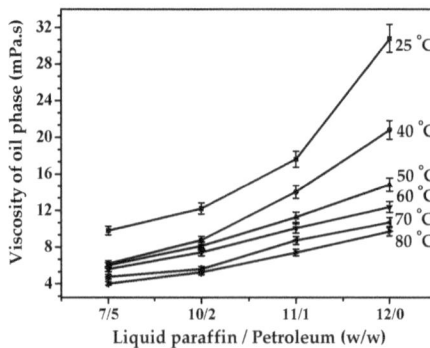

Figure 8. Viscosities of oil phase with different weight ratios of liquid paraffin to petroleum ether.

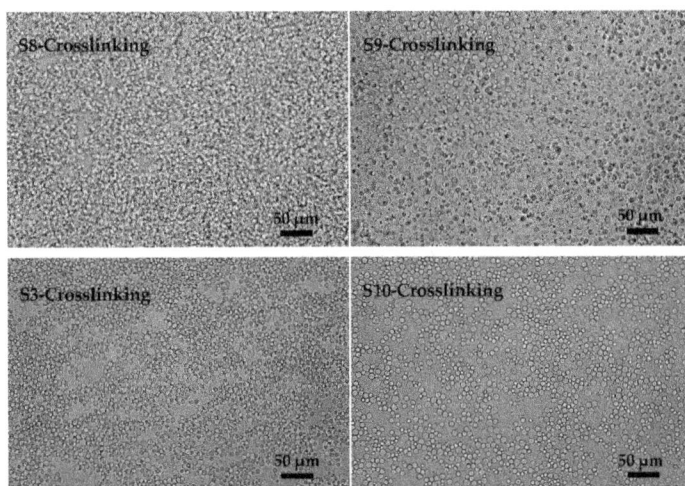

Figure 9. Microscopic photographs of KGM microspheres prepared with different oil phase compositions (7:5, 10:2, 11:1 and 12:0, LP/EP, *w/w*): S8-Crosslinking, S9-Crosslinking, S3-Crosslinking and S10-Crosslinking were the crosslinking results of **S8**, **S9**, **S3** and **S10**.

3.4. Effects of the Feeding Methods of EC on the Preparation of KGM Microspheres

In the crosslinking process, flocculation and coalescence occurred frequently. Figure 10 shows the evolution of emulsions with time in the absence of a crosslinking agent. Flocculation and aggregation occurred at different times in the emulsions with different stabilities. The results indicated that low temperature contributed to the stability of the emulsions. An increase in the temperature increased the solubility of the emulsifier in the oil and aqueous phases, so that part of the emulsifier would be desorbed from the oil-water interface, leading to the reduction of the interfacial film strength and the decline of the emulsion stability [32]. Meanwhile, the viscosities of the oil phase and aqueous phase increased with temperature (Figure 11a). The collision energy between the emulsion droplets would decrease for the high motion resistance resulting from the high viscosity of the oil phase. The high viscosity of the aqueous phase could partly inhibit flocculation due to the high viscoelasticity. In addition, high temperature would promote the Brown movement and increase the chance of collision between droplets which would induce the flocculation and coalescence of the emulsions. Although the interfacial tension got larger at low temperature (Figure 11b), it played a minor role. Temperature had an important influence on the stability of the emulsions.

Before coalescence and flocculation occurred, emulsions generally kept stable for a period. During this period, increasing the crosslinking rate would partly inhibit the coalescence and flocculation. So it was very important to increase the crosslinking rate at the beginning of the crosslinking process. Here, we adjusted the crosslinking rate by changing the feeding method of the crosslinking agent (EC). In the traditional method, the EC was added slowly at the crosslinking temperature (60 °C). Here, two different methods were proposed. In the first one, EC was added at 40 °C and then the temperature was increased gradually to 60 °C. Sufficient dispersion of EC in the oil phase could be achieved at 40 °C. During this period the emulsions were in stable state because that low temperature contributed to emulsion stability. As the crosslinking temperature (60 °C) was reached, the well-dispersed EC would react more rapidly with the emulsion droplets. In the second one, EC solution (20 wt %, EC was mixed with oil phase adequately) was added in one batch at the reaction temperature (60 °C). It was easier for the EC solution to spread to the microsphere surface. It can be seen from the results in Figure 12 that flocculation and coalescence occurred in S5-Crosslinking 1 which was cured by the traditional method. In S5-Crosslinking 2, emulsions were stirred at 60 °C for

1 h after emulsification and then EC was added. In this process, emulsion coalescence was almost fully developed before crosslinking and the new emulsions were in stable state, so the obtained microspheres had better dispersity than those in S5-Crosslinking 1. Both of the methods (Figure 12 (S5-Crosslinking 3 and S5-Crosslinking 4)) improved the crosslinking results. The final microspheres prepared with the two different feeding methods had better dispersity and the coalescence and flocculation were effectively suppressed.

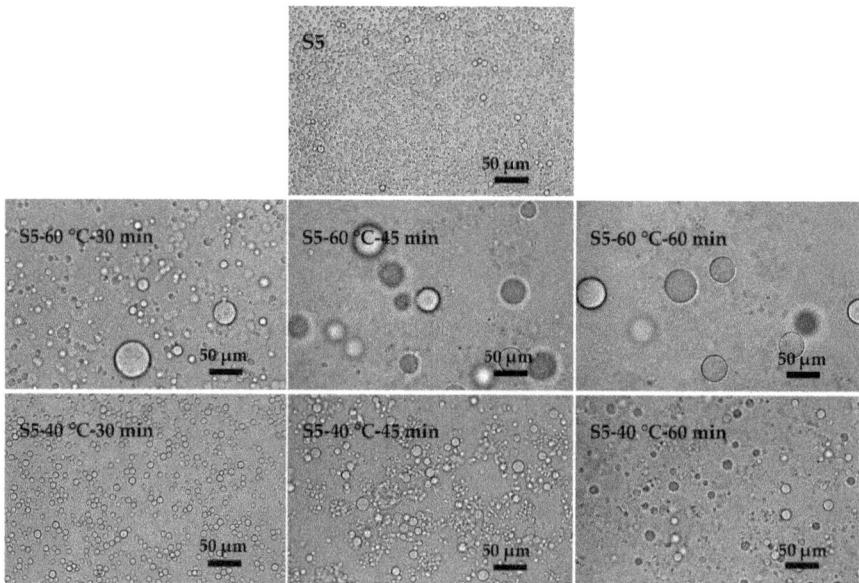

Figure 10. Microscopic photographs of KGM emulsions for **S5**. After emulsification, emulsions of **S5** were kept at 60 °C (or 40 °C) for 30, 45 and 60 min with stirring.

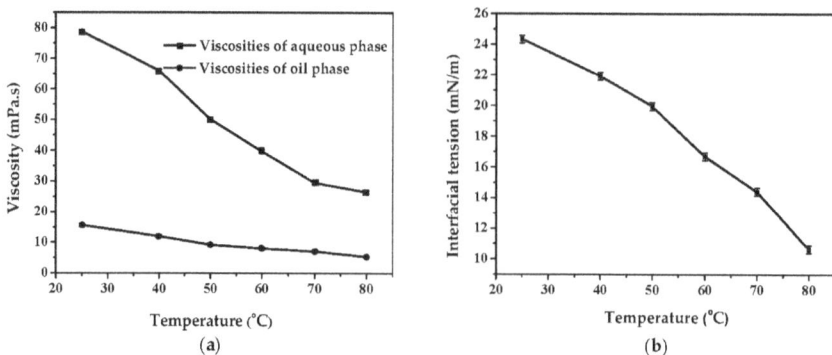

Figure 11. The viscosities of oil (LP/EP, 11/1 (*w*/*w*)) and aqueous phases (KGM I) and the interfacial tension between oil phase and aqueous phases at different temperatures. (**a**) Viscosities of oil and aqueous phases at different temperatures; (**b**) Interfacial tensions between oil and aqueous phases at different temperatures.

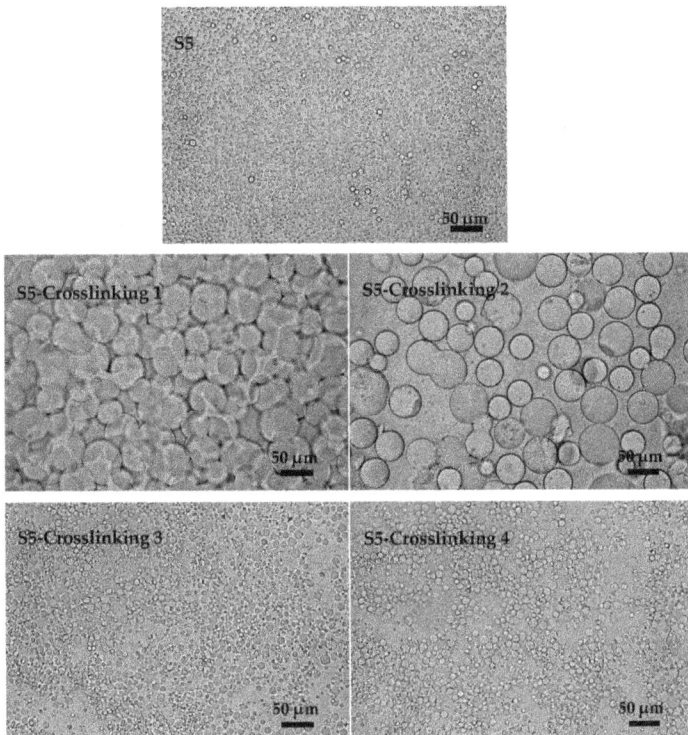

Figure 12. Microscopic photographs of KGM microspheres prepared by different feeding modes of EC: **S5** was KGM emulsions before crosslinking and S5-Crosslinking 1 (EC was added slowly at 60 °C), S5-Crosslinking 2 (emulsions were stirred at 60 °C for 1 h and then EC was added), S5-Crosslinking 3 (EC solution (20 wt %, EC was mixed with oil phase adequately) was added in one batch at 60 °C) and S5-Crosslinking 4 (EC was added at 40 °C and then the temperature was increased to 60 °C) were the crosslinking results of **S5**.

4. Conclusions

Uniform small-sized KGM microspheres were successfully prepared with the EP membrane we developed by premix membrane emulsification. The stability of emulsions had an important influence on the uniformity and dispersity of the final KGM microspheres. It was found that the preparation condition of the emulsifier 5 wt % PO-5s, KGM III (145.6 mPa·s), and weight ratio of LP to PE 11:1 was in favor of the uniformity and dispersity of the final KGM microspheres. The addition of emulsifier was beneficial to the stability of emulsions, but the excess emulsifier would increase the hydrophilicity of the membrane and produce large droplets. High viscosity of the KGM solution could partly improve the uniformity and stability of emulsions and the greater the oil phase viscosity was, the better the emulsion stability was. In addition, the feeding methods of EC played a significant role in the crosslinking results. Different from the traditional feeding method of EC, the two feeding methods proposed in this study could partly inhibit the coalescence and flocculation, which effectively improved the crosslinking results.

Acknowledgments: We would like to thank the financial support from the National Natural Science Foundation of China (No.21336010), Beijing Natural Science Foundation (No.2162013), National Key Technology R&D Program of the Ministry of Science and Technology (No. 2013 BAB01B03) and National High Technology Research and Development Program of China (863 Program, No.2014AA021006).

Polymers **2016**, *8*, 53

Author Contributions: Guanghui Ma suggested and supervised the work and performed the article editing. Zhiguo Su provided lots of constructive suggestions about this work. Yace Mi and Weiqing Zhou designed and performed the experiments and analyzed the data. Juan Li provided significant guidance on the preparation of KGM microspheres. Rongyue Zhang contributed to the design of EP membrane.

Conflicts of Interest: The authors declare no conflict of interest.

Abbreviations

The following abbreviations are used in this manuscript:

EP	Epoxy-based Polymer membrane
KGM	Konjac Glucomannan
SPG	Shirasu Porous Glass membrane
LP	Liquid Paraffin
PE	Petroleum Ester
CV	Coefficient of Variation
EC	Epoxy Chloropropane
Po-5s	Hexagly Cerinpenta Ester

References

1. Vanderbeek, P.B.; Fasano, C.; O'Malley, G.; Hornstein, J. Esophageal obstruction from a hygroscopic pharmacobezoar containing glucomannan. *Clin. Toxicol.* **2007**, *45*, 80–82. [CrossRef]
2. Han1A, B.; Zhang, C.; Luo, X. Study of konjac glucomannan esterification with dicarboxylic anhydride and effect of degree of esterification on water absorbency. *Key Eng. Mater.* **2011**, *501*, 42–46. [CrossRef]
3. Zhang, T.; Xue, Y.; Li, Z.; Wang, Y.; Xue, C. Effects of deacetylation of konjac glucomannan on alaska pollock surimi gels subjected to high-temperature (120 °C) treatment. *Food Hydrocoll.* **2015**, *43*, 125–131. [CrossRef]
4. Wang, S.; Zhan, Y.; Wu, X.; Ye, T.; Li, Y.; Wang, L.; Chen, Y.; Li, B. Dissolution and rheological behavior of deacetylated konjac glucomannan in urea aqueous solution. *Carbohydr. Polymer* **2014**, *101*, 499–504. [CrossRef] [PubMed]
5. Tian, D.; Li, S.; Liu, X.; Wang, J.; Liu, C. Synthesis and properties of konjac glucomannan-*graft*-poly(acrylic acid-*co*-trimethylallyl ammonium chloride) as a novel polyampholytic superabsorbent. *Adv. Polym. Technol.* **2013**, *32*, E131–E140. [CrossRef]
6. Shen, C.; Li, W.; Zhang, L.; Wan, C.; Gao, S. Synthesis of cyanoethyl konjac glucomannan and its liquid crystalline behavior in an ionic liquid. *J. Polym. Res.* **2012**, *19*, 1–8. [CrossRef]
7. Xiong, Z.D.; Zhou, W.Q.; Sun, L.J.; Li, X.N.; Zhao, D.W.; Chen, Y.; Li, J.; Ma, G.H.; Su, Z.G. Konjac glucomannan microspheres for low-cost desalting of protein solution. *Carbohydr. Polym.* **2014**, *111*, 56–62. [CrossRef] [PubMed]
8. Wang, J. Study on konjac glucomannan accumulation in cell suspension culture of *Amorphophallus konjac*. *J. Anhui Agric. Sci.* **2010**, *27*, 14836–14838.
9. Guo, X.M.; Wang, G.L.; Wei-Lin, Y.E.; Zhou, Z.X.; Xiang, Y.; Zhu, X. Hydrogen peroxide biosensor based on immobilizing enzyme with deacetyled konjac glucomannan. *J. Instrum. Anal.* **2008**, *27*, 581–580.
10. Chen, L.G.; Liu, Z.L.; Zhuo, R.X. Synthesis and properties of degradable hydrogels of konjac glucomannan grafted acrylic acid for colon-specific drug delivery. *Polymer* **2005**, *46*, 6274–6281. [CrossRef]
11. Korkiatithaweechai, S.; Umsarika, P.; Praphairaksit, N.; Muangsin, N. Controlled release of diclofenac from matrix polymer of chitosan and oxidized konjac glucomannan. *Marine Drugs* **2011**, *9*, 1649–1663. [CrossRef] [PubMed]
12. Ma, G.H.; Su, Z.G.; Wang, J.X.; Ge, J.L. A konjac Glucomannan Microsphere and Its Preparation Method. CN 101113180 B, 8 December 2010.
13. Shen, J. A Method to Prepare Konjac Glucomanan Microspheres. CN 102627779 A, 8 August 2012.
14. Nakashima, T.; Shimizu, M.; Kukizaki, M.; Nakashima, T.; Shimizu, M.; Kukizaki, M. Membrane emulsification by microporous glass. *Key Eng. Mater.* **1992**, *61*, 513–516. [CrossRef]
15. Piacentini, E.; Drioli, E.; Giorno, L. Membrane emulsification technology: Twenty-five years of inventions and research through patent survey. *J. Membr. Sci.* **2014**, *468*, 410–422. [CrossRef]

16. Van der Graaf, S.; Schroen, C.; Boom, R.M. Preparation of double emulsions by membrane emulsification—A review. *J. Membr. Sci.* **2005**, *251*, 7–15. [CrossRef]

17. Zhang, Y.Q.; Xie, B.J.; Gan, X. Advance in the applications of konjac glucomannan and its derivatives. *Carbohydr. Polym.* **2005**, *60*, 27–31. [CrossRef]

18. Charcosset, C.; Limayem, I.; Fessi, H. The membrane emulsification process—A review. *J. Chem. Technol. Biotechnol.* **2004**, *79*, 209–218. [CrossRef]

19. Kohler, J.; Chase, D.B.; Farlee, R.D.; Vega, A.J.; Kirkland, J.J. Comprehensive characterization of some silica-based stationary phases for high-performance liquid chromatofraphy. *J. Chromatogr.* **1986**, *352*, 275–305. [CrossRef]

20. Mi, Y.; Zhou, W.; Li, Q.; Gong, F.; Zhang, R.; Ma, G.; Su, Z. Preparation of water-in-oil emulsions using a hydrophobic polymer membrane with 3D bicontinuous skeleton structure. *J. Membr. Sci.* **2015**, *490*, 113–119. [CrossRef]

21. Vladisavljevic, G.T.; Kobayashi, I.; Nakajima, M.; Williams, R.A.; Shimizu, M.; Nakashima, T. Shirasu porous glass membrane emulsification: Characterisation of membrane structure by high-resolution X-ray microtomography and microscopic observation of droplet formation in real time. *J. Membr. Sci.* **2007**, *302*, 243–253. [CrossRef]

22. Bardenhagen, I.; Dreher, W.; Fenske, D.; Wittstock, A.; Bäumer, M. Fluid distribution and pore wettability of monolithic carbon xerogels measured by ^{1}H NMR relaxation. *Carbon* **2014**, *68*, 542–552. [CrossRef]

23. Mi, Y.; Zhou, W.Q.; Li, Q.; Zhang, D.L.; Zhang, R.Y.; Ma, G.H.; Su, Z.G. Detailed exploration of structure formation of an epoxy-based monolith with three-dimensional bicontinuous structure. *RSC Adv.* **2015**, *5*, 55419–55427. [CrossRef]

24. Miyagawa, Y.; Katsuki, K.; Matsuno, R.; Adachi, S. Effect of oil droplet size on activation energy for coalescence of oil droplets in an *o/w* emulsion. *Biosci. Biotechnol. Biochem.* **2015**, 1–3. [CrossRef] [PubMed]

25. Becher, P. *Encyclopedia of Emulsion Technology, Vol. 1: Basic Theory*; Marcel Dekker: New York, NY, USA, 1983; p. 320.

26. Kabalnov, A.S.; Shchukin, E.D. Ostwald ripening theory—Applications to fluorocarbon emulsion stability. *Adv. Colloid Interface Sci.* **1992**, *38*, 69–97. [CrossRef]

27. Schröder, V.; Behrend, O.; Schubert, H. Effect of dynamic interfacial tension on the emulsification process using microporous, ceramic membranes. *J. Colloid Interface Sci.* **1998**, *202*, 334–340. [CrossRef]

28. Zhou, Q.; Wang, L.; Ma, G.; Su, Z. Multi-stage premix membrane emulsification for preparation of agarose microbeads with uniform size. *J. Membr. Sci.* **2008**, *322*, 98–104. [CrossRef]

29. Silva, H.D.; Cerqueira, M.A.; Vicente, A.A. Influence of surfactant and processing conditions in the stability of oil-in-water nanoemulsions. *J. Food Eng.* **2015**, *167*, 89–98. [CrossRef]

30. Wasan, D.T. *Interfacial Transport Processes and Rheology*; Award Lecture on Chemical Engineering Education Spring 1992; Chicago, IL, USA, 1992.

31. Sjöblom, J. *Emulsions and Emulsion Stability*; Taylor & Francis: New York, NY, USA, 2006.

32. Wang, X.; Brandvik, A.; Alvarado, V. Probing interfacial water-in-crude oil emulsion stability controls using electrorheology. *Energy Fuels* **2010**, *24*, 6359–6365. [CrossRef]

polymers

MDPI

Article

Initiator Systems Effect on Particle Coagulation and Particle Size Distribution in One-Step Emulsion Polymerization of Styrene

Baijun Liu [1], Yajun Wang [1], Mingyao Zhang [1,*] and Huixuan Zhang [1,2]

[1] School of Chemical Engineering, Changchun University of Technology, Changchun 130012, China; liubaijun111@126.com (B.L.); yjwang_bio@163.com (Y.W.); zhanghx@mail.ccut.edu.cn (H.Z.)

[2] Changchun Institute of Applied Chemistry, Chinese Academy of Sciences, Changchun 130022, China

* Correspondence: zmy@mail.ccut.edu.cn; Tel.: +86-431-8571-6465

Academic Editor: Haruma Kawaguchi
Received: 6 December 2015; Accepted: 15 February 2016; Published: 19 February 2016

Abstract: Particle coagulation is a facile approach to produce large-scale polymer latex particles. This approach has been widely used in academic and industrial research owing to its higher polymerization rate and one-step polymerization process. Our work was motivated to control the extent (or time) of particle coagulation. Depending on reaction parameters, particle coagulation is also able to produce narrowly dispersed latex particles. In this study, a series of experiments were performed to investigate the role of the initiator system in determining particle coagulation and particle size distribution. Under the optimal initiation conditions, such as cationic initiator systems or higher reaction temperature, the time of particle coagulation would be advanced to particle nucleation period, leading to the narrowly dispersed polymer latex particles. By using a combination of the Smoluchowski equation and the electrostatic stability theory, the relationship between the particle size distribution and particle coagulation was established: the earlier the particle coagulation, the narrower the particle size distribution, while the larger the extent of particle coagulation, the larger the average particle size. Combined with the results of previous studies, a systematic method controlling the particle size distribution in the presence of particle coagulation was developed.

Keywords: emulsion polymerization; particle size distribution; coagulation; initiator

1. Introduction

Emulsion polymerization is a widely used process for the production of rubber, plastic, coating, and adhesives in industry [1–5]. Control over size and polydispersity in these applications is required because of the close relationship between the properties of the polymer and the particle size distribution [6–13]. Thus, how to control particle size and polydispersity has gradually become an essential issue. Until today, many technologies based on emulsion polymerization including seeded emulsion polymerization and agglomeration method [1,3,6,10,14–16] have been proposed to control the particle size and distribution. Among these technologies, the particle coagulation technology has been accepted as a highly effective approach to prepare nanoparticles in both industrial production and theoretical investigation [1,7,15,17–19].

Particle coagulation is a process that occurs in the period of the particle nucleation and growth. Even though it is induced by the increase in the interfacial energy change, it is also a kinetically controlled process. Particle coagulation can be divided into two periods, as shown in Scheme 1: (1) the process of the particle aggregation is determined by the probability of the particle collision. Some factors such as the viscosity of the media and the reaction temperature can directly affect this process [20]. In agreement with the Smoluchowski equation [21,22], the kinetic of the particle coagulation can be expressed by $-\dfrac{dN}{dt} = k_c N_0^2$, where $-\dfrac{dN}{dt}$ is the particle coagulation rate, and k_c and N_0 are a constant and particle number, respectively. In another period, several particles merge into a larger one, which is determined by the particle structure and glass transition temperature [23,24]. The effect of simple reaction parameters on particle coagulation in emulsion polymerization using pure water as solvent has been well illustrated by many researchers. For instance, Dobrowolska *et al.* investigated the effect of ionic strength on the extent of particle coagulation and particle size distribution, and illustrated that higher ionic strength could decrease the thickness of the diffuse electric double layer, and further increase the extent of particle coagulation [17,18,25]. Chern *et al.* elaborated the role of surfactant systems in determining particle coagulation and found that the relationship between the particle number and the surfactant concentration [26,27].

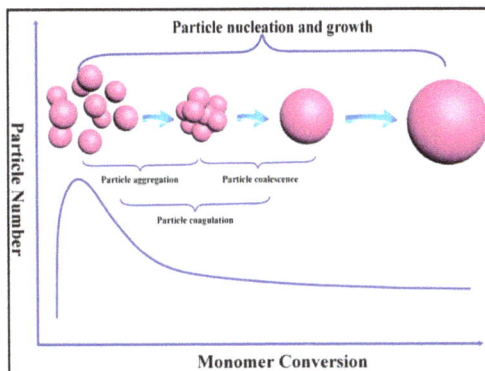

Scheme 1. The schematic diagram of particle coagulation in one-step emulsion polymerization.

As usual, the emulsion polymerization is carried out in water as the medium. However, recent investigations indicated that the addition of co-solvent to the medium played an important role in determining the particle size and distribution of the ultima latex. For example, Adelnia *et al.* investigated that the effect methanol on the characteristics of the Poly (methacrylate-*co*-butyl acrylate) latex, and found that the addition of methanol increased the medium viscosity and further facilitated the particle coagulation [20]. Kim *et al.* also carried out the investigation of soap-free emulsion polymerization of methyl methacrylate in different methanol solutions, and indicated that the polymerization product and behavior resembled those typical of pure water [28].

In our previous reports, we carried out a series of experimental investigations about particle coagulation, and stressed the role of a co-solvent such as methanol in determining the extent of the particle coagulation [29–32]. With the increase in methanol concentration in aqueous phase, the interfacial tension between the aqueous phase and particles decreased. This decreased the adsorption capability of the surfactant molecules on particle surface and further decreased the repulsive potential energy. On the other hand, the attractive potential increased because of the decrease in the Hamaker constant. As a result, the extent of the particle coagulation was enhanced. In addition, the addition of the methanol also increased the length of the critical chain length (CCL) when polymer chains precipitated from the aqueous phase, further increasing the initial particle size, and decreasing the polymerization reaction rate. Even though particle coagulation has been studied extensively, some fundamental issues such as the effect of the initiator systems on particle coagulation are still puzzling to investigators.

Radical polymerization initiator systems can be divided into many categories. From the solubility view, the initiator systems can be divided into water-soluble and oil-soluble initiators; from another view, it can also be divided into ionic and nonionic types. The ionic type initiator system is further divided into cationic and anionic types based on the difference in the charges. The effect of the initiator system on particle coagulation seems to be an easy problem, but it is indeed an extremely complicated phenomenon. According to the Smith–Ewart theory, pure water was usually chosen as solvent, and the particle number increased with increasing initiator concentration [3], but the Smoluchowski equation indicated that the rate of particle coagulation also increased with increasing particle number, as described above [21,22]. Thus, the addition of the initiator seems to play an opposite role in determining particle number, and what is its main function? The ionic initiator not only plays the role of the initiator, but also function as an electrolyte, which increases the complexity of the particle coagulation. In the case of opposite charged surfactant and initiator, does the shielding action between surfactant molecules adsorbed on particle surface and the initiator chain ends affect the extent of the particle coagulation? Can the time of particle coagulation be adjusted by initiator systems?

In regard to the effect of the oil-soluble initiator on particle nucleation mechanism in emulsion polymerization, two main mechanisms for the production of radicals were postulated. One of the mechanisms considered that the first radicals were generated in the monomer-swollen polymer particles/monomer droplets/monomer swollen micelles, and desorbed to the continuous phase; another mechanism considered that the radical formed in the continuous phase were derived from the fraction of the oil-soluble initiator dissolved in the continuous phase [33–37]. Nomura *et al.* indicated the relationship between the particle number (and molecular weight) and recipe compositions when oil-soluble initiator was used in pure water solvent. For example, Nomura *et al.* carried out the unseeded and seeded emulsion polymerization of styrene using azodiisobutyronitrile (AIBN) as an oil-soluble initiator and concluded that the latex particles were formed in the emulsifier micelles, and the particle number was proportional to the 0.30th power of the concentration of the initiator [38,39]. Capek *et al.* also reported that the kinetic aspects initiated by AIBN in the presence of a blend of anionic and non-ionic surfactant conditions and stressed that the addition of AIBN decreased the molecular weight and the polymerization reaction rate [40,41]. Recently, Gugliotta *et al.* investigated the role of the oil-soluble initiator in governing the particle nucleation in mini-emulsion polymerization, and indicated that the oil-soluble initiator could promote droplet nucleation and control the second nucleation [42]. Even though the kinetic model and experimental evidence of the emulsion polymerization initiated by oil-soluble initiator have been developed, it is still difficult to control the particle size distribution of ultima latex when oil-soluble initiator is used, especially to prepare narrowly dispersed polymer latex particles.

To address these problems, in this study, different initiator systems such as water-soluble potassium persulfate (an anionic type initiator), oil-soluble azodiisobutyronitrile (a nonionic type), and 2,2'-azobis [2-methylpropionamidine] dihydrochloride (AIBA) (a cationic ionic type) were chosen to initiate the polymerization reaction of styrene in the presence of methanol solution. The evolution of the particle size, number, and distribution as a function of polymerization time was traced for achieving a comprehensive understanding about particle coagulation. The polymerization condition selected here is based on the idea that obvious particle coagulation occurs, although, in many situations, particle coagulation is not obvious. The effect of reaction parameters except initiator systems on the particle coagulation and particle size distribution could be obtained from our previous studies [23,29–32].

2. Materials and Methods

Chemical Styrene (St, 99%), supplied by the Shanghai Chemical Reagent Corporation (Shanghai, China), was distilled under vacuum to remove the inhibitors prior to polymerization and used as the monomer. Sodium dodecyl sulfate (SDS; 99%; Aladdin, Shanghai, China) and Potassium carbonate (K_2CO_3; 98.5%; Aladdin, Shanghai, China) were used as the surfactant and electrolyte without any further purification. Potassium persulfate (KPS; 99.5%; Aladdin, Shanghai, China), Azodiisobutyronitrile (AIBN; 98%; Aladdin, Shanghai, China), 2,2'-azobis [2-methylpropionamidine] dihydrochloride (AIBA, 99%; Aladdin, Shanghai, China) were used as the initiator. Double distilled-deionized (DDI) water was used in all experiments. The polymerization fundamental recipe for investigating the effect of initiator systems on particle coagulation is shown in Table 1.

Table 1. The polymerization fundamental recipe for investigating the effect of initiator systems on particle coagulation.

Formulation parameter	Reagents	Quantity (g)
Monomer	St	100
Surfactant	SDS	1.5
Initiator	KPS/AIBN/AIBA	Variable
Electrolyte	K_2CO_3	0.6
Co-solvent	methanol	30

2.1. Polymerization Reactions

The emulsion polymerization reactions of styrene were carried out using a 500 mL glass reactor equipped with four necks for string mechanically with an anchor stirrer (Bar Length: 300 mm; Blade Diamete: 45 mm; Surface Coating Material: Polytetrafluoroethylene), condensing the reflux with cold water, purging with nitrogen gas, and sampling an aliquot of the solution with a pipette. The SDS, K_2CO_3, DDI and co-solvent (methanol) were added to reaction equipment according to this sequence; subsequently, monomer was also added when all auxiliaries were dissolved in the aqueous phase. Nitrogen (N_2) purging was carried out for 30 min before the initiator was added to the reactor. When the initiator dissolved in some DDI or monomer was added into the equipment, the polymerization reaction begins. The polymerization reaction was carried out under the N_2 atmosphere, the reaction temperature and stirring rate were set as 65 °C and 250 rpm, respectively. During the polymerization process, 2–3 g latex was withdrawn from the reactor using a syringe at appropriate intervals to analyze the monomer conversion, particle size distribution and particle number. The polymerization reaction time was set as 6 h.

2.2. Characterization

The particle size and distribution was measured by the Brookhaven 90plus Particle Size Analyzer (Brookhaven, NY, USA) and transmission electron microscopy (TEM) (JEOL 1210, Tokyo, Japan). The polydispersity index (PDI) was directly obtained from the Particle Size Analyzer, and defined by ISO 13321: 1996 E. The particle number (N_p) was obtained by the following equation [32]:

$$N_p = \frac{6XM_0}{\pi d_p^3 \rho_p} \tag{1}$$

where M_0 is the monomer concentration in the unit mass aqueous phase (1 kg for aqueous phase), ρ_p is the polymer density, X is the monomer fractional conversion (which could be obtained by the gravimetric method) and d_p is the average size of the latex particle.

3. Results

3.1. KPS Initiator System

To better understand the effect of the initiator system on the particle coagulation and particle size distribution, the KPS system was first considered. The KPS initiator system is one of the most widely used initiators in conventional emulsion polymerization in pure water solvent [3,10]. The theory of Smith–Ewart (micellar theory) predicted a proportionality value of 0.40 between particle number and KPS initiator concentration; Sajjadi *et al.* performed the emulsion polymerization of butyl acrylate using KPS as initiator according to the method of Capek [41], and found that the particle number was proportional to 0.39th power of KPS concentration [43]. These theories confirmed the relationship between the particle size and initiator concentration, indicated that the particle size of the ultimate latex decreased with increasing the initiator concentration. In this study, the particle size of the ultimate latex corresponding to different initiator concentrations is shown in Figure 1. The particle size increased with increasing initiator concentrations (97.9, 107.5, 136.7, and 161.5 nm at 0.3, 0.6, 0.9 and 1.2 wt %, respectively), and was in reverse trend to that of particle size in the conventional emulsion polymerization.

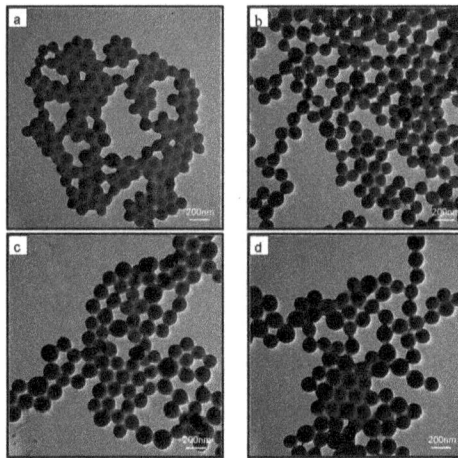

Figure 1. TEM micrographs of the final particles prepared by emulsion polymerization of styrene using methanol solution (20/80 w/w) as the polymerization medium and various amount of initiator KPS: (**a**) 0.3 wt %; (**b**) 0.6 wt %; (**c**) 0.9 wt %; and (**d**) 1.2 wt %.

In addition, the evolution of monomer conversion, particle size and number as a function of the polymerization process for different initiator concentrations is traced, as shown in Figure 2. The curve of monomer conversion *vs.* the polymerization time, as shown in Figure 2a, indicates that the polymerization rate using 0.3 wt % KPS initiator system was the fastest among these reactions and was opposite to the results reported by Carro *et al* [21]. Figure 2b shows that the particle size rapidly reached 90 nm at 0.2 conversion in the presence of 1.2 wt % KPS. Meanwhile, the particle size corresponding to the systems consisting of 0.3, 0.6, and 0.9 wt % KPS only attained 54.2, 62.1, and 70.0 nm, respectively. Figure 2c shows that the particle number increased with increasing conversion, then decreased until reaching a constant particle number at appropriate conversion. The decrease in particle number and the increase in particle size indicated that the particle coagulation occurred during the polymerization process [29–32]. With increasing initiator concentration, the extent of particle coagulation gradually increased and the time of particle coagulation advanced. In addition, the evolution of the system temperature inside reactor also reflects some information about the particle coagulation, as shown in Figure 2d. The increase in the system temperature was mainly attributed to the heat released from the system larger than the heat transmission from the reaction system to the environment [44,45]. The increase in temperature was also considered as an evaluated parameter to indicate the polymerization reaction rate [46]. Figure 2d shows that the maximum system temperature decreased with increasing initiator concentrations from 71.2, 70.2, and 69.1 °C at 0.3, 0.6 and 0.9 wt % KPS, respectively, to 67.9 °C at 1.2 wt %. Moreover, the time corresponding to the maximum system temperature also advanced as the initiator KPS concentrations increased. With increasing initiator concentrations, the initial reaction rate was enhanced, and the particle number also increased. However, by increasing the particle number, the extent and time of the particle coagulation were enhanced. This decreased the rate of the polymerization reaction and led to the decrease of the system temperature [20].

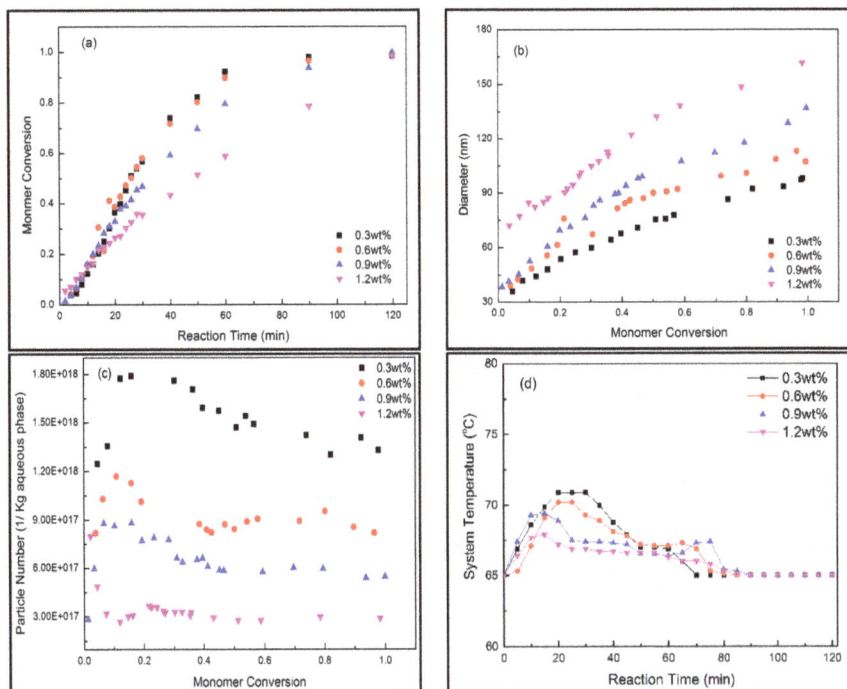

Figure 2. Plots of monomer conversion (**a**); particle size (**b**); number (**c**); and system temperature (**d**) against reaction time (or monomer conversion) for the emulsion polymerization of styrene using methanol solution (20/80 w/w) as the polymerization medium and various amount of initiator KPS ([I] = 0.3, 0.6, 0.9 and 1.2 wt %).

As a result of early particle coagulation, the width of particle size distribution of the final latex particles decreased with increasing initiator concentrations, as shown in Figure 3. The polydispersity index of ultimate latex particles decreased from 0.036, 0.030, and 0.024 at 0.3, 0.6, and 0.9 wt % KPS, respectively, to 0.016 at 1.2 wt % KPS. The polydispersity index of ⩽0.1 indicates that the ultimate latex particles had a relatively narrow particle size distribution [47]. Meanwhile, the particles obtained by particle coagulation were spherical and smooth, indicating that narrowly dispersed latex particles could be prepared in the presence of particle coagulation.

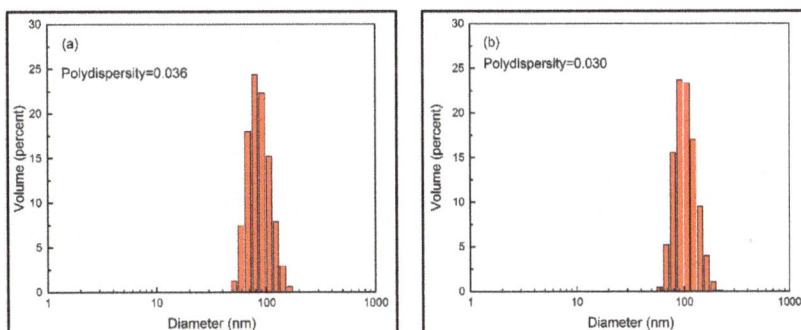

Figure 3. *Cont.*

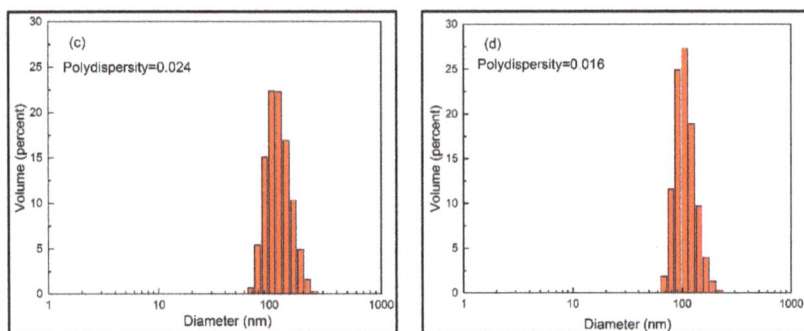

Figure 3. Particle size distribution of final latex particles prepared by emulsion polymerization of styrene using methanol solution (20/80 w/w) as the polymerization medium and various amount of initiator KPS: (**a**) 0.3 wt %; (**b**) 0.6 wt %; (**c**) 0.9 wt %; and (**d**) 1.2 wt %.

To ensure the role of increasing initiator concentration, the polymerization reactions with 0.6 wt % KPS initiator were carried out at the initiation reaction temperatures of 55 and 75 °C. The initiator decomposition rate is well known to increase with increasing reaction temperature [48]. Therefore, increasing initiation temperature also increased the primary radicals, thus increasing the extent of particle coagulation [45,46,48]. The TEM images of the ultimate latex particles prepared at different reaction temperature are shown in Figure 4. As expected, the average size of latex particles increased with increasing reaction temperature from 91.2 to 111.2 nm. Moreover, the polydispersity index value decreased from 0.031 at 55 °C to 0.015 at 75 °C.

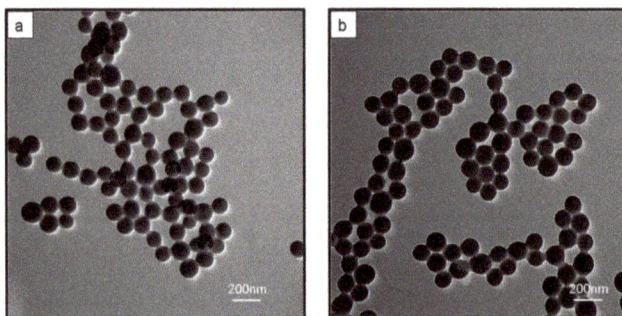

Figure 4. TEM micrographs of the final particles prepared by emulsion polymerization of styrene using methanol solution (20/80 w/w) as the polymerization medium and various reaction temperatures: (**a**) 55 °C and (**b**) 75 °C. The TEM micrograph of the final particles prepared by emulsion polymerization of styrene at 65 °C is shown in Figure 1b.

The evolution of the monomer conversion, average particle size, number and system temperature as a function of reaction time (or monomer conversion) of KPS system at different initiation reaction temperatures is shown in Figure 5. The rapid increase in average particle size and the decrease in the particle number confirmed the particle coagulation behavior during the polymerization process. The initial time of the decrease in the particle number showed that the starting time of particle coagulation advanced to the nucleation period with increasing initiation reaction temperature. As the reaction temperature increased, the total polymerization reaction time shortened, as shown in Figure 5a. In addition, the time corresponding to the highest system temperature shifted to 10 min from 24 min with increasing reaction temperature, as expected.

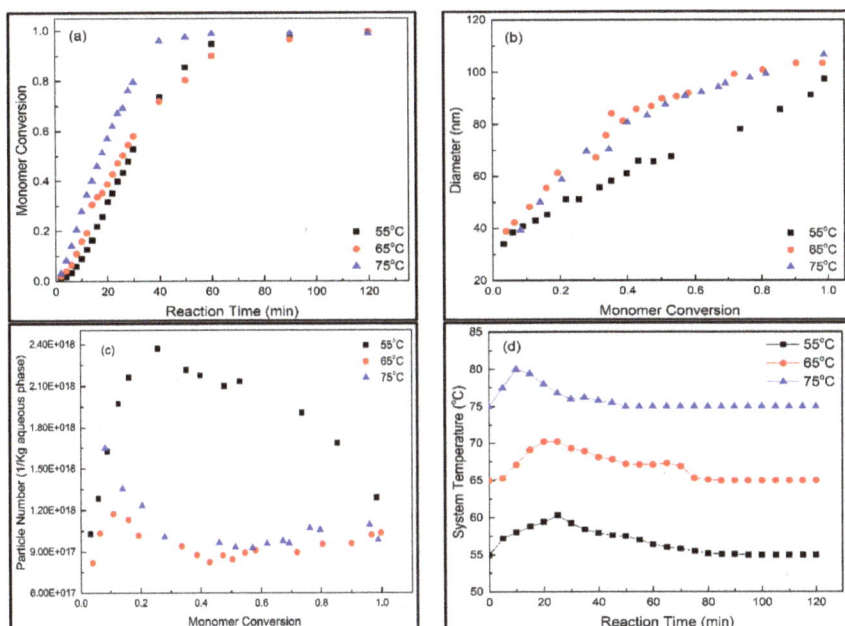

Figure 5. Plots of monomer conversion (**a**); particle size (**b**); number (**c**); and system temperature (**d**) against reaction time (or monomer conversion) for the emulsion polymerization of styrene using methanol solution (20/80 *w/w*) as the polymerization medium and various reaction temperatures (55, 65, and 75 °C).

3.2. AIBN and AIBA Initiator Systems

As mentioned above, other type of initiators, such as oil-soluble AIBN and cationic AIBA, could also be used to initiate emulsion polymerization reactions. Differing from the conventional KPS initiator, the oil-soluble initiator AIBN scarcely dissolves in aqueous media during the polymerization process. From the AIBN solubility view, the initiator decomposition reaction should be carried out in these monomer droplets or swollen micelles according to a similar bulk method. However, the small reaction volume of the swollen-micelles (or monomer droplets) made the initiator radicals easy to recombine, further limiting the initiation reactions that occurred in monomer droplets [33–37]. Thus, the kinetic model of emulsion polymerization using oil-soluble AIBN was closer to the zero-one model, rather than the pseudo-bulk model [48–50]. Figure 6 shows that the TEM images of ultimate latex particles prepared using AIBN and AIBA initiators at 0.6 wt % concentration. The results indicated that the latex particles using AIBN and AIBA were much larger than those using KPS initiator system. The average particle size attained were 141.2 and 224.6 nm with AIBN and AIBA, respectively.

The curves of monomer conversion, particle size, particle number, and system temperature *vs.* monomer conversion (or reaction time) shown in Figure 7 indicate that particle coagulation was scarcely observed for the AIBN system because no decrease in particle number was observed in Figure 7c. In contrast to AIBN system, an obvious particle coagulation process was observed in the AIBA system, as shown in Figure 7c. Figure 7a shows that the KPS initiator system only needed ~120 min reaction time when the monomer conversion reached 1, even though particle coagulation occurred. However, the systems initiated by AIBN and AIBA needed 240 and 300 min, respectively. The curves of the particle number *vs.* monomer conversion indicate that the initial particle numbers of the AIBA and the AIBN systems were much smaller than the KPS one. The evolution of the system temperature inside the reactor initiated by AIBA and AIBN is also traced, as shown in Figure 7d. However, the variation in the system temperature was difficult to observe because of the slow polymerization rate. In the whole polymerization process, the maximum system temperatures for AIBA and AIBN were slightly higher than the reaction temperature, 67.5 and 67.2 °C, respectively.

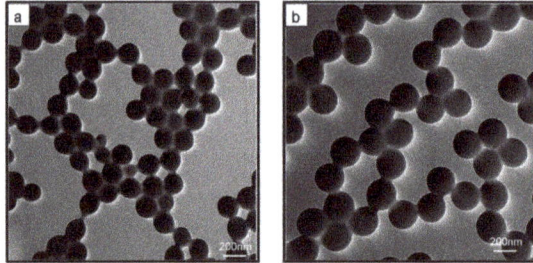

Figure 6. TEM micrographs of the final particles prepared by emulsion polymerization of styrene using methanol solution (20/80 w/w) as the polymerization medium and various initiator systems: (**a**) AIBN and (**b**) AIBA. The TEM micrograph of the final particles prepared by emulsion polymerization of styrene initiated by KPS is shown in Figure 1b.

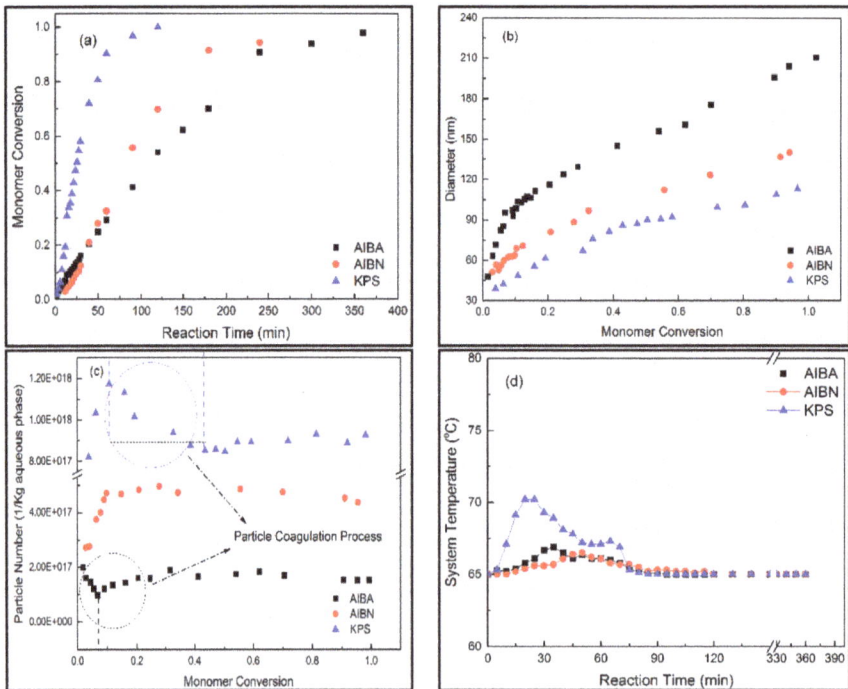

Figure 7. Plots of the monomer conversion (**a**); particle size (**b**); particle number (**c**); and system temperature (**d**) against monomer conversion (or reaction time) for the emulsion polymerization of styrene using methanol solution (20/80 w/w) as the polymerization medium and various types of initiator.

4. Discussion

Based on these experimental results, the effect of initiator systems on particle coagulation and particle size distribution is discussed.

The extent of particle coagulation was indeed controlled by adjusting the initiator system, especially by using ionic initiators such as KPS and AIBA. The addition of KPS not only increased the primary radical concentration in aqueous media, but also increased the ionic strength. The increase in the primary radical concentration increased the probability of particle collision, and further promoted particle coagulation occurred. This process could be expressed by the modified Von Smoluchowski equation [51]:

$$B\,(i,\,j) = f\,(\gamma)\,N\,(i)\,N\,(j)\,\big[d_p\,(i) + d_p\,(j)\big]^3 \tag{2}$$

where $B(i,j)$ and $f(\gamma)$ are the number of collisions between particles i and j class and constant, respectively. Because of the increase in the primary particle number, the $B(i,j)$ directly increased, thus enhancing the extent of particle coagulation. This process was controlled by the kinetics, as described in the Introduction Section. As the particle coagulation was carried out, $N(i)$ and $N(j)$ decreased, thus the collision frequency between particles i and j class decreased. Meanwhile, the larger particles obtained by the particle coagulation between particles i and j class was difficult to aggregate unless the number of larger particles attained a critical value. On the other hand, KPS and AIBA initiator could also be considered as the electrolyte in polymerization recipe, and the addition of ionic initiator enhanced the ionic strength. Furthermore, the thickness of electrical double layer surrounding the particle surface was compressed, which decreased the particle stability and increased the extent of particle coagulation [25]. The oil-soluble AIBN initiator was not ionic, therefore, the addition of AIBN scarcely affected the thickness of electrical double layer, and, furthermore, particle coagulation was not obvious. Meanwhile, when the oil-soluble AIBN was used as initiator, the vast majority of the initiator dissolved in monomer droplets, monomer-swollen micelles, and polymer particles, and only a small quantity of initiator dissolved in the aqueous phase, which initiated the polymerization reaction. Because the pairs of radicals produced within a volume as small as a monomer-swollen latex particle or a monomer-swollen micelle are easily recombined, the free radicals produced from the fraction of initiator dissolved in the aqueous phase are responsible for particle formation and growth. Therefore, the deactivation of the oil-soluble initiator plays a significant role in determining the initial particle number and the rate of the polymerization reaction. Nomura *et al.* indicated that the efficiency of the oil-soluble initiator was only 1/9 of that of KPS in pure water solvent [52]. In the oil-soluble initiator system, the relationship between particle number and initiator was similar to conventional emulsion initiated by KPS system, indicating that the particle number decreased with decreasing initiator concentration. The deactivation of the AIBN is equivalent to the decrease in initiator concentration. As a result, the initial particle number of the oil-soluble initiator was much smaller than that initiated by KPS. The smaller the particle number was, the smaller the frequency of particle collision achieved, which limited the occurrence of particle coagulation. Notably, in the AIBA initiator system, the initial particle number was much smaller than that of the KPS system. This might be attributed to the *in situ* charge neutralization. The polymer initiated AIBA possessed cationic chain ends, which shielded the anionic charge of surfactant molecules SDS adsorbed on the particle surface. As a result, particle coagulation occurred in the early polymerization period. Even though particle coagulation occurred in particle nucleation period, the slight increase in particle number was also observed in this period (Figure 7c). The increase in particle number might be attributed to the new particle formed in aqueous phase. The particle number would increase when the rate of particle formed in aqueous phase is larger than the rate of particle coagulation [30,32]. As the polymerization reaction was carried out, the particle number gradually attained a constant value because a balance between particle nucleation and coagulation was obtained. The earlier the particle coagulation, the narrower the particle size distribution of ultimate latex particles. The curves of particle number *vs.* monomer conversion confirm that the time of particle coagulation could affect the polydispersity of the final latex particles because of the competitive growth mechanism [53]. If the particle coagulation occurred in the early nucleation period, the latex particles were completely swollen by styrene monomer because of the presence of monomer droplets. Thus, the particles were practically soft, leading to the primary coagulation of latex particles easily merging into the larger one [25]. The process of particle coagulation was controlled by the kinetic factors such as the particle collision. In the next particle growth period, the larger particle size had a smaller ratio of surface and volume and, furthermore, had a relatively slow rate of capturing monomer radicals, resulting in a much slower growth rate of larger particles than the smaller ones [6]. As a result, the narrowly dispersed latex particles were obtained in the presence of particle coagulation.

From the thermodynamic view, the initiator systems also affect the electrostatic force among the particles themselves. For instance, the cationic AIBA initiator systems decomposed into initiator radicals, and the initiator radicals reacted with styrene monomer dissolved in aqueous media and formed the monomer radicals with cationic initiator ends. These monomer radicals were captured by the swollen micelles formed by SDS molecules, and further shielded the negative charge of the surfactant SDS molecules adsorbed on the particle surface, and decreased the electrostatic repulsion among the particle themselves. As a result, *in situ* charge neutralization process occurred, further increasing the extent of the particle coagulation, and promoting particle coagulation that occurred in the early nucleation period.

5. Conclusions

In conclusion, we demonstrated the effect of initiator systems on the particle coagulation and particle size distribution of ultimate latex particles. The change in the initiator systems not only affects the extent of the particle coagulation, but also determines the time of particle coagulation. In the anionic KPS system, with the increase in the initiator concentration, the extent of particle coagulation was enhanced, resulting in obtaining larger size and narrowly dispersed latex particles. The cationic AIBA initiator played a significant role in determining the time of particle coagulation. The positive charges derived from AIBA chain ends shield the negative charge of surfactant SDS molecules adsorbed on the particle surface, leading to *in situ* charge neutralization, further enhancing the extent of particle coagulation and advancing the time of particle coagulation. The kinetics of oil-soluble AIBN systems seems similar to that of a conventional polymerization because of the deactivation of the oil-soluble initiator. Thus, combined with our previous reports on the particle coagulation, a systematic method for controlling the particle size distribution in the presence of particle coagulation was achieved.

Acknowledgments: The authors appreciate the financial support from the National Natural Scientific Foundation of China (No. 51573022).

Author Contributions: Baijun Liu and Mingyao Zhang conceived and designed the experiments; Yajun Wang performed the experiments; Baijun Liu and Huixuan Zhang analyzed the data; Mingyao Zhang contributed reagents/materials/analysis tools; Baijun Liu wrote the paper.

Conflicts of Interest: The authors declare no conflict of interest.

References

1. Feiz, S.; Navarchian, A.H. Emulsion polymerization of styrene: Simulation the effects of mixed ionic and non-ionic surfactant system in the presence of coagulation. *Chem. Eng. Sci.* **2012**, *69*, 431–439. [CrossRef]
2. Shen, K.; Wang, Y.; Ying, G.; Liang, M.; Li, Y. Poly (styrene–isoprene–butadiene-*g*-SAN) graft copolymers: Size-controllable synthesis and their toughening properties. *Colloids Surf. A* **2015**, *467*, 216–223. [CrossRef]
3. Chern, C.S. Emulsion polymerization mechanisms and kinetics. *Prog. Polym. Sci.* **2006**, *31*, 443–486. [CrossRef]
4. An, L.; Di, Z.; Yu, B.; Pu, J.; Li, Z. The Effect of allylic sulfide-mediated irreversible addition-fragment chain transfer on the emulsion polymerization kinetics of styrene. *Polymers* **2015**, *7*, 1918–1938. [CrossRef]
5. Chen, L.; Wu, F. Preparation and characterization of pure polyacrylate polymer colloid through emulsion polymerization using a novel initiator. *Colloids Surf. A* **2011**, *392*, 300–304. [CrossRef]

6. Zhang, Q.; Han, Y.; Wang, W.; Song, T.; Chang, J. A theoretical and experimental investigation of the size distribution of polystyrene microspheres by seeded polymerization. *J. Colloid Interface Sci.* **2010**, *342*, 62–67. [CrossRef] [PubMed]

7. Telford, A.M.; Pham, B.T.; Neto, C.; Hawkett, B.S. Micron-sized polystyrene particles by surfactant-free emulsion polymerization in air: Synthesis and mechanism. *J. Polym. Sci. Polym. Chem.* **2013**, *51*, 3997–4002. [CrossRef]

8. Hamberger, A.; Landfester, K. Influence of size and functionality of polymeric nanoparticles on the adsorption behavior of sodium dodecyl sulfate as detected by isothermal titration calorimetry. *Colloid Polym. Sci.* **2011**, *289*, 3–14. [CrossRef]

9. Mu, Y.; Qiu, T.; Li, X. Monodisperse and multilayer core–shell latex via surface cross-linking emulsion polymerization. *Mater. Lett.* **2009**, *63*, 1614–1617. [CrossRef]

10. Rao, J.P.; Geckeler, K.E. Polymer nanoparticles: Preparation techniques and size-control parameters. *Prog. Polym. Sci.* **2011**, *36*, 887–913. [CrossRef]

11. Shibuya, K.; Nagao, D.; Ishii, H.; Konno, M. Advanced soap-free emulsion polymerization for highly pure, micron-sized, monodisperse polymer particles. *Polymer* **2014**, *55*, 535–539. [CrossRef]

12. Sajjadi, S. Extending the limits of emulsifier-free emulsion polymerization to achieve small uniform particles. *RSC Adv.* **2015**, *5*, 58549–58560. [CrossRef]

13. Ishii, H.; Kuwasaki, N.; Nagao, D.; Konno, M. Environmentally adaptable pathway to emulsion polymerization for monodisperse polymer nanoparticle synthesis. *Polymer* **2015**, *77*, 64–69. [CrossRef]

14. Cho, K.D.; Jang, K.; Yang, G.H.; Chang, J.G.; Ha, K.R.; Park, J.W.; Song, S.M. Synthesis and properties of agglomerating agent for high-solids NBR latices. *J. Appl. Polym. Sci.* **2002**, *84*, 276–282. [CrossRef]

15. Ito, F.; Ma, G.; Nagai, M.; Omi, S. Study of particle growth by seeded emulsion polymerization accompanied by electrostatic coagulation. *Colloids Surf. A* **2002**, *201*, 131–142. [CrossRef]

16. Kim, J.W.; Suh, K.D. Monodisperse micro-sized polystyrene particles by seeded polymerization: Effect of seed crosslinking on monomer swelling and particle morphology. *Polymer* **2000**, *41*, 6181–6188. [CrossRef]

17. Peach, S. Coagulative nucleation in surfactant-free emulsion polymerization. *Macromolecules* **1998**, *31*, 3372–3373. [CrossRef]

18. Ito, F.; Makino, K.; Ohshima, H.; Terada, H.; Omi, S. Salt effects on controlled coagulation in emulsion polymerization. *Colloids Surf. A* **2004**, *233*, 171–179. [CrossRef]

19. Kemmere, M.F.; Meuldijk, J.; Drinkenburg, A.A.H.; German, A.L. Aspects of coagulation during emulsion polymerization of styrene and vinyl acetate. *J. Appl. Polym. Sci.* **1998**, *69*, 2409–2421. [CrossRef]

20. Adelnia, H.; Pourmahdian, S. Soap-free emulsion polymerization of poly(methyl methacrylate-*co*-butyl acrylate): Effects of anionic comonomers and methanol on the different characteristics of the latexes. *Colloid Polymer Sci.* **2014**, *292*, 197–205. [CrossRef] [PubMed]

21. Carro, S.; Herrera-Ordonez, J.; Castillo-Tejas, J. On the evolution of the rate of polymerization, number and size distribution of particles in styrene emulsion polymerization above CMC. *J. Polym. Sci. Polym. Chem.* **2010**, *48*, 3152–3160. [CrossRef]

22. Von Smoluchowski, M. Investigation of a mathematical theory on the coagulation of colloidal suspensions. *Z. Physik. Chem.* **1917**, *92*, 129–168.

23. Liu, B.; Zhang, M.; Ao, Y.; Zhang, H. Crosslinking network structure effects on particle coagulation in the emulsion polymerization of styrene in methanol solution. *Colloid Polym. Sci.* **2015**, *293*, 1577–1581. [CrossRef]

24. Feng, Y.; Huang, S.; Teng, F. Controlled particle size and synthesizing mechanism of microsphere of poly(MMA-BuMA) prepared by emulsion polymerization. *Polym. J.* **2009**, *41*, 266–271. [CrossRef]

25. Dobrowolska, M.E.; Koper, G.J.M. Optimal ionic strength for nonionically initiated polymerization. *Soft Matter* **2014**, *10*, 1151–1154. [CrossRef] [PubMed]

26. Chern, C.S.; Lin, S.Y.; Chen, L.J. Emulsion polymerization of styrene stabilized by mixed anionic and nonionic surfactants. *Polymer* **1997**, *38*, 1977–1984. [CrossRef]

27. Krishnan, S.; Klein, A.; El-Aasser, M.S.; Sudol, E.D. Effect of surfactant concentration on particle nucleation in emulsion polymerization of *n*-butyl methacrylate. *Macromolecules* **2003**, *36*, 3152–3159. [CrossRef]

28. Kim, G.; Lim, S.; Lee, B.H.; Shim, S.E.; Choe, S. Effect of homogeneity of methanol/water/monomer mixture on the mode of polymerization of MMA: Soap-free emulsion polymerization versus dispersion polymerization. *Polymer* **2010**, *51*, 1197–1205. [CrossRef]

29. Liu, B.; Zhang, M.; Cheng, H.; Fu, Z.; Zhou, T.; Chi, H.; Zhang, H. Large-scale and narrow dispersed latex formation in batch emulsion polymerization of styrene in methanol–water solution. *Colloid Polym. Sci.* **2014**, *292*, 519–525. [CrossRef]

30. Liu, B.; Zhang, M.; Gui, Y.; Chen, D.; Zhang, H. Effect of aqueous phase composition on particle coagulation behavior in batch emulsion polymerization of styrene. *Colloids Surf. A* **2014**, *452*, 159–164. [CrossRef]

31. Liu, B.; Zhang, M.; Zhou, C.; Fu, Z.; Wu, G.; Zhang, H. Hydrophilicity of polymer effects on controlled particle coagulation in batch emulsion polymerization. *Colloid Polym. Sci.* **2014**, *292*, 1347–1353. [CrossRef]

32. Liu, B.; Sun, S.; Zhang, M.; Ren, L.; Zhang, H. Facile synthesis of large scale and narrow particle size distribution polymer particles via control particle coagulation during one-step emulsion polymerization. *Colloids Surf. A* **2015**, *484*, 81–88. [CrossRef]

33. Asua, J.M.; Rodriguez, V.S.; Sudol, E.D.; El-Aasser, M.S. The free radical distribution in emulsion polymerization using oil-soluble initiators. *J. Polym. Sci. Part A* **1989**, *27*, 3569–3587. [CrossRef]

34. Alduncin, J.A.; Forcada, J.; Barandiaran, M.J.; Asua, J.M. On the main locus of radical formation in emulsion polymerization initiated by oil-soluble initiators. *J. Polym. Sci. Part A* **1991**, *29*, 1265–1270. [CrossRef]

35. Nomura, M.; Ikoma, J.; Fujita, K. Kinetics and mechanisms of emulsion polymerization initiated by oil-soluble initiators. IV. Kinetic modeling of unseeded emulsion polymerization of styrene initiated by 2,2'-azobisisobutyronitrile. *J. Polym. Sci. Part A* **1993**, *31*, 2103–2113. [CrossRef]

36. Autran, C.; de La Cal, J.C.; Asua, J.M. (Mini) emulsion polymerization kinetics using oil-soluble initiators. *Macromolecules* **2007**, *40*, 6233–6238. [CrossRef]

37. Luo, Y.; Schork, F.J. Emulsion and miniemulsion polymerizations with an oil-soluble initiator in the presence and absence of an aqueous-phase radical scavenger. *J. Polym. Sci. Part A* **2002**, *40*, 3200–3211. [CrossRef]

38. Nomura, M.; Tobita, H.; Suzuki, K. Emulsion polymerization: Kinetic and mechanistic aspects. *Adv. Polym. Sci.* **2005**, *175*, 1–128.

39. Nomura, M.; Yamada, A.; Fujita, S.; Sugimoto, A.; Ikoma, J.; Fujita, K. Kinetics and mechanism of emulsion polymerization initiated by oil-soluble initiators. II. Kinetic behavior of styrene emulsion polymerization initiated by 2,2'-azoisobutyronitrile. *J. Polym. Sci. Part A* **1991**, *29*, 987–994. [CrossRef]

40. Capek, I. On the role of oil-soluble initiators in the radical polymerization of micellar systems. *Adv. Colloid Inter. Sci.* **2001**, *91*, 295–334. [CrossRef]

41. Capek, I.I. Emulsion polymerization of butyl acrylate IV. Effects of initiator type and concentration. *Polym. J.* **1994**, *26*, 1154–1162. [CrossRef]

42. Ronco, L.I.; Minari, R.J.; Gugliotta, L.M. Particle nucleation using different initiators in the miniemulsion polymerization of styrene. *Braz. J. Chem. Eng.* **2015**, *32*, 191–199. [CrossRef]

43. Sajjadi, S.; Brooks, B.W. Butyl acrylate batch emulsion polymerization in the presence of sodium lauryl sulphate and potassium persulphate. *J. Polym. Sci. Polym. Chem.* **1999**, *37*, 3957–3972. [CrossRef]

44. Cutting, G.R.; Tabner, B.J. Reaction temperature profiles and radical concentration measurements on batch emulsion copolymerization of methyl methacrylate and butyl acrylate. *Eur. Polym. J.* **1995**, *12*, 1215–1219. [CrossRef]

45. Capek, I.; Potisk, P. Microemulsion and emulsion polymerization of butyl acrylate-I. Effect of the initiator type and temperature. *Eur. Polym. J.* **1995**, *31*, 1269–1277. [CrossRef]

46. Liu, B.; Zhang, M.; Chen, D.; Liu, S.; Han, Y.; Zhang, H. Exothermal behavior and particle scale evolution in high solid content one-step batch emulsion polymerization. *J. Dispersion Sci. Technol.* **2015**, *36*, 205–212. [CrossRef]

47. Camli, S.T.; Buyukserin, F.; Balci, O.; Budak, G.G. Size controlled synthesis of sub-100nm monodisperse poly(methylmethacrylate) nanoparticles using surfactant-free emulsion polymerization. *J. Colloid Interface Sci.* **2010**, *344*, 528–532. [CrossRef] [PubMed]

48. Gardon, J.L. Emulsion polymerization. I. Recalculation and extension of the Smith-Ewart theory. *J. Polym. Sci. Polym. Chem.* **1968**, *6*, 623–641. [CrossRef]

49. Gardon, J.L. Emulsion polymerization. II. Review of experimental data in the context of the revised Smith-Ewart theory. *J. Polym. Sci. Polym. Chem.* **1968**, *6*, 643–664. [CrossRef]

50. Yamamoto, T. Synthesis of micron-sized polymeric particles in soap-free emulsion polymerization using oil-soluble initiators and electrolytes. *Colloid Polym. Sci.* **2012**, *290*, 1023–1031. [CrossRef]

51. Mayer, M.J.J.; Meuldijk, J.; Thoenes, D. Dynamic modeling of limited particle coagulation in emulsion polymerization. *J. Appl. Polym. Sci.* **1996**, *59*, 83–90. [CrossRef]

52. Nomura, M.; Suzuki, K. A new kinetic interpretation of the styrene microemulsion polymerization. *Macromol. Chem. Phys.* **1997**, *198*, 3025–3039. [CrossRef]

53. Li, Z.; Cheng, H.; Han, C.C. Mechanism of narrowly dispersed latex formation in a surfactant-free emulsion polymerization of styrene in acetone-water mixture. *Macromolecules* **2012**, *45*, 3231–3299. [CrossRef]

MDPI AG

St. Alban-Anlage 66

4052 Basel, Switzerland

Tel. +41 61 683 77 34

Fax +41 61 302 89 18

http://www.mdpi.com

Polymers Editorial Office

E-mail: polymers@mdpi.com

http://www.mdpi.com/journal/polymers

www.ingramcontent.com/pod-product-compliance
Lightning Source LLC
Chambersburg PA
CBHW041216220326
41597CB00033BA/5986